Microbiology of Fresh Produce

Emerging Issues in Food Safety

SERIES EDITOR, Michael P. Doyle

Microbiology of Fresh Produce
Edited by Karl R. Matthews

ALSO IN THIS SERIES

Microbial Source Tracking (2006)
Edited by Jorge W. Santo Domingo and Michael J. Sadowsky

Microbial Risk Analysis in Foods (2007)
Edited by Donald W. Schaffner

***Enterobacter sakazakii* (2007)**
Edited by Jeffrey M. Farber and Stephen Forsythe

Microbiology of Fresh Produce

EDITED BY

Karl R. Matthews

Department of Food Science, Cook College,
Rutgers, The State University of New Jersey,
New Brunswick, New Jersey

WASHINGTON, D.C.

Address editorial correspondence to ASM Press, 1752 N St. NW,
Washington, DC 20036-2904, USA

Send orders to ASM Press, P.O. Box 605, Herndon, VA 20172, USA
Phone: 800-546-2416; 703-661-1593
Fax: 703-661-1501
E-mail: books@asmusa.org
Online: estore.asm.org

Library of Congress Cataloging-in-Publication Data

Microbiology of fresh produce / edited by Karl R. Matthews.
p. cm—(Emerging issues in food safety)
Includes bibliographical references and index.
ISBN 1-55581-357-7
1. Fruit—Microbiology. 2. Vegetables—Microbiology. I. Matthews,
Karl R. II. Series.
QR115.M494 2006
664'.001'579—dc22

2005053611

10 9 8 7 6 5 4 3 2 1

Cover illustration: Confocal micrographs of fluorescently labeled bacterial cells in association with a root (upper left), leaf vein (upper right), stomata (lower left), and leaf tip (lower right). Courtesy of Maria Brandl, Produce and Microbiology Research Unit, USDA, ARS, WRRC—Albany, California.

Contents

Contributors

ARVIND A. BHAGWAT
Produce Quality and Safety Laboratory, Henry A. Wallace Beltsville Agricultural Research Center, Agricultural Research Service, U.S. Department of Agriculture, 10300 Baltimore Avenue, Bldg. 002, BARC-W, Beltsville, MD 20705-2350

ELIZABETH A. BIHN
Department of Food Science, Cornell University, Ithaca, NY 14853

MARIA T. BRANDL
Produce Safety and Microbiology Research Unit, Western Regional Research Center, Agricultural Research Service, U.S. Department of Agriculture, Albany, CA 94710

CHRISTINE M. BRUHN
Center for Consumer Research, University of California, Davis, Davis, CA 95616-8598

WILLIAM F. FETT
Food Safety Intervention Technologies Research Unit, Eastern Regional Research Center, Agricultural Research Service, U.S. Department of Agriculture, Wyndmoor, PA 19038

JORGE M. FONSECA
Yuma Agricultural Center, The University of Arizona, Yuma, AZ 85364

TONG-JEN FU
National Center for Food Safety and Technology, U.S. Food and Drug Administration, Summit-Argo, IL 60501

ROBERT B. GRAVANI
Department of Food Science, Cornell University, Ithaca, NY 14853

ROBERT E. MANDRELL
Produce Safety and Microbiology Research Unit, Western Regional Research Center, Agricultural Research Service, U.S. Department of Agriculture, Albany, CA 94710

KARL R. MATTHEWS
Department of Food Science, Cook College, Rutgers, The State University of New Jersey, New Brunswick, NJ 08901

ETHAN B. SOLOMON
DuPont Chemical Solutions Enterprise, Experimental Station Laboratory—E402/3223, P.O. Box 80402, Wilmington, DE 19880-0402

MARY LOU TORTORELLO
National Center for Food Safety and Technology, U.S. Food and Drug Administration, Summit-Argo, IL 60501

Series Editor's Foreword

This is the first book in a new series of monographs that will address emerging topics of the microbiological safety of foods. Reports of estimates of tens of millions of cases of foodborne illness in the United States annually indicate the magnitude of this problem. Epidemiologists report that in the United States foodborne illnesses are more common than influenza or the common cold. Microorganisms or their metabolites are responsible for the vast majority of foodborne illnesses, but there are many unknowns regarding disease agents, including their epidemiology, mechanisms of pathogenicity, infectious or toxic dose, host sensitivity, detection and subtyping methods, and treatments for inactivation. Microbiologists continue to unravel these mysteries, often renouncing age-old beliefs that were long considered dogma. Hence, the need exists to provide scientists interested in the microbiological safety and quality of foods a medium for an authoritative presentation of seminal issues of major significance to the field.

Fresh produce consumption in the United States is increasing at a dramatic rate, more than doubling over the past two to three decades. Concomitant with this has been a substantial increase in the importation of fruits and vegetables and in the incidence of foodborne illnesses associated with fresh produce consumption. An incredible amount of new information has been published during the past few years addressing the microbiological safety of fresh produce. In some instances, questions have been raised regarding long-held dogmas, an example being that the internal contents of intact fruits and vegetables are sterile. Recent reports have dispelled this principle with experimental findings revealing that tomatoes and lettuce can be internally contaminated by harmful microorganisms, depending on growing and processing conditions. Therefore, it is most fitting that the first monograph in this series focuses on this important and timely topic.

My compliments to Karl Matthews and his team of coauthors who have been truly remarkable in pulling together in record time the state-of-the-art information regarding the microbiological safety of fresh produce. Well done.

MICHAEL P. DOYLE, Series Editor
Emerging Issues in Food Safety

Preface

The microbiological quality of fresh fruits and vegetables is essential to ensuring the availability of a high-quality, safe product for the consumer. Consumption of fresh fruits and vegetables is an important part of a healthy diet, supplying much-needed vitamins, minerals, and fiber. Health promotion aspects of fruits and vegetables are beginning to be widely acknowledged through their role in the prevention of heart disease, cancer, and diabetes. Consumers now expect fresh produce year-round and in the United States purchase grapes, cantaloupe, or lettuce grown in Central or South America within days of harvest.

Preventing the exposure of raw fruits and vegetables to microorganisms while in the field, orchard, or vineyard is impossible. Fruit and vegetables are produced in a natural environment and are therefore exposed to a wide range of microorganisms. The majority of microorganisms that are recovered from raw fruits and vegetables at harvest do not represent a human health risk but may cause spoilage. However, the number of outbreaks caused by foodborne pathogens associated with fresh produce has increased during the past three decades. Without further efforts to understand the complex interactions between microbes and fresh fruits and vegetables and the mechanisms by which contamination occurs from the farm to the fork, this trend will likely continue.

Consumers now demand fresh fruits and vegetables that may have been produced in remote areas of the world packaged for convenience and available at a reasonable price at the local supermarket. Produce must be of high quality microbiologically, or the product will spoil or potentially cause human illness. Knowledge of the microbiology of fresh fruits and vegetables preharvest and at all stages postharvest (i.e., processing, packaging, storage, and at retail outlets) is imperative to the development of new technologies and implementation of innovative methods to ensure that a wholesome, microbiologically safe product reaches the consumer.

This book provides the essential information on the microbiology of fresh produce. It focuses on the unique challenges to the control of microorganisms on produce from the farm to the consumer. The latest technologies for reducing microbial load, packaging, and detection are discussed. Consumer knowledge of produce handling, foodborne illness risks, and future product desires are covered. The association of human pathogens with outbreaks of foodborne illness and a perspective on the microbiological safety of produce are presented. This book will be of interest to microbiologists, food safety experts, extension specialists, food scientists, and specialists in academia, government, and industry concerned with the microbiological quality of fresh fruits and vegetables.

I am grateful to each of the subject experts who authored chapters of this book and to the many people who have made important contributions to it. Their promptness and cooperation made possible the timely completion of this book.

KARL R. MATTHEWS

Microbiology of Fresh Produce
Edited by Karl R. Matthews

Microorganisms Associated with Fruits and Vegetables

1

Karl R. Matthews

Consumption of fresh fruits and vegetables is integral to a healthy diet. The U.S. Department of Agriculture recommends five to nine servings of fruits and vegetables daily (http://www.nal.usda.gov/fnic/Fpyr/pyramid.html). The World Health Organization and the Food and Agriculture Organization recommend the intake of a minimum of 400 g of fruits and vegetables per day. Fruits and vegetables supply much-needed vitamins, minerals, and fiber. Produce can play an important role in health through the prevention of heart disease, cancer, and diabetes. Moreover, consumption of fresh fruits and vegetables is a key component of programs designed to address the international epidemic of overweightness and obesity (http://www.who.int/dietphysicalactivity/fruit).

In the United States, the per capita consumption of fresh fruits and vegetables increased from 254 pounds in 1980 to 328 pounds in 2000. This equates to a 19 and a 29% increase in per capita consumption of fresh fruits and vegetables, respectively (10). The trend in increased consumption of fruits and vegetables is expected to continue through 2020, with fruit consumption increasing by 24 to 27% and vegetable consumption increasing by 19 to 24% (38). Increased consumption of fruits and vegetables may have unintended consequences. Since fruit and vegetables are produced in a natural environment, they are vulnerable to contamination by human pathogens. Approximately 12% of foodborne illnesses in the 1990s were linked to fresh produce items (23).

KARL R. MATTHEWS, Department of Food Science, Cook College, Rutgers, The State University of New Jersey, New Brunswick, NJ 08901.

MICROBES ASSOCIATED WITH FRESH FRUITS AND VEGETABLES

Numerous factors can influence contamination of produce, including the use of manure as a fertilizer; contaminated agricultural water; contaminated harvesting equipment; hygiene practices of workers in the field, packinghouse, and processing plant; and the presence of feral animals in fields and packinghouses. Since produce is consumed raw and no intervention practices are employed that can effectively control or eliminate pathogens prior to consumption, it is a potential source of foodborne illness. Challenges exist in all countries, developing and developed, with respect to devising and implementing measures to control and prevent contamination of produce with pathogens that present a human health risk. The U.S. Food and Drug Administration (FDA) considers the safety of imported and domestic fruits and vegetables a priority and has initiated a number of activities to ensure that consumers receive microbiologically safe produce (4). In 1999 and 2000, the FDA initiated surveys of imported and domestic fresh produce, respectively (http://www.cfsan.fda.gov/~dms/prodsur6.html; http://www.cfsan.fda.gov/~dms/prodsu10.html). The focus was on high-volume products including cantaloupe, celery, cilantro, and loose-leaf lettuce. Among domestic produce, 11 (1%) of 1,028 samples were positive for either *Salmonella* or *Shigella* spp. and no items were positive for *Escherichia coli* O157:H7. Similar results were obtained for imported produce: 44 (4.4%) of 1,003 samples were positive for either *Shigella* or *Salmonella* spp., and no samples were positive for *E. coli* O157:H7.

The types of pathogens screened for in the FDA studies of domestic and imported produce were limited. Specific etiological agents linked with produce-associated outbreaks are many. Bacteria (*Salmonella* and *Shigella* spp., *E. coli* O157:H7, *Listeria monocytogenes, Campylobacter jejuni,* and *Yersinia enterocolitica*), viruses (calicivirus, hepatitis A virus, and norovirus), parasites (*Cyclospora cayetanensis, Cryptosporidium parvum*, and *Giardia lamblia*), helminths, and a range of other microbes have been linked to cases of human illness following consumption of contaminated produce (5, 50, 51, 54). Factors influencing the presence, numbers, and source of microbes associated with produce prior to harvest include, but are not limited to, the type of produce, weather conditions, and agronomic practices.

POTENTIAL SOURCES OF PATHOGENS

Preharvest

Perhaps the two most important sources of pathogens prior to harvest are water used for irrigation and for the application of insecticides and fungicides, and manure applied as a fertilizer. A number of studies have demonstrated the long-term survival of *E. coli* O157:H7 and *Salmonella* in manure. *E. coli* O157:H7 survived for 21 months in manure that had been collected from experimentally inoculated sheep and that was held outside under fluctuating environmental conditions (37). The researchers in this study also reported the survival of *E. coli* O157:H7 in bovine manure for 47 days. *Salmonella enterica* serovar Typhimurium and *E. coli* O157:H7 survived from 6 days to 3 weeks in bovine manure and from 2 days to 5 weeks in bovine manure slurry (28). The long-term survival of *Salmonella* in animal feces is not surprising since the issue has been studied for several decades (18, 46, 56). The long-term survival of *E. coli* O157:H7 in well, reservoir, recreational, and municipal waters is also well documented (9, 62). Many routes exist for the introduction of contaminated manure into farm water sources, and once present, *E. coli* O157:H7 can survive for extended periods in waters that may ultimately be applied to growing crops (32). Further research has shown that the method of application of irrigation water directly influences whether the organism can be found associated with the edible portion of the plant at harvest (52). Spray irrigation (i.e., water contacts the surface of the plant) is used in the cultivation of approximately 50% of the lettuce grown in the United States (http://www.nass.usda.gov/census/census97/fris/fris.htm). Spray irrigation resulted in a greater number of lettuce plants' testing positive for *E. coli* O157:H7 at harvest following a single exposure to the target pathogen (52). A study conducted in Nigeria showed that crops (lettuce, carrots, and amaranth) were positive for *Salmonella*, *Vibrio* spp., or *E. coli* following irrigation with water that also tested positive for the same pathogens (42). The waters used for irrigation were contaminated with wastes from residential houses and commercial establishments. The irrigation practices and microbiological quality of irrigation water used in foreign countries could potentially impact human illness in the United States through consumption of imported contaminated fresh fruits and vegetables. These results confirm the suggestion by the FDA that "the quality of water in direct contact with the edible portion of produce may need to be of better (microbiological) quality compared to uses where there is minimal contact" (3).

Irrigation water and manure fertilizer often top the list of sources of pathogens in the field; however, other relevant sources include migratory birds and feral animals. *E. coli* O157:H7 has been isolated from goose feces collected from fields used for production of lettuce. Deer feces have also tested positive for *E. coli* O157:H7. Feral animals have also been shown to carry *Salmonella*, *Campylobacter*, and other human bacterial pathogens (36). Contamination of fresh fruits and vegetables with viruses is likely the result of contamination by workers during harvest and not contaminated water, manure fertilizer, or feral animals. The majority of microorganisms that are recovered from raw fruits and vegetables at harvest do not represent a human health risk but may cause spoilage. Pseudomonads, yeasts, molds, and *Bacillus, Erwinia, Xanthomonas,* and *Clostridium* spp. reflect the microflora present in the orchard, field, and vineyard at harvest. Spoilage microbes degrade the tissue of the product, providing a niche for pathogenic microorganisms to reside. Controlling spoilage microbes on fruits and vegetables will also aid in the control of human pathogens (60). Ultimately, spoilage microbes may impact survival of human pathogens on raw fruits and vegetables.

Some controversy exists concerning whether there are differences in the microbiological qualities of organic versus conventionally grown produce. However, very few studies have addressed the question of the microbiological quality of organically and conventionally grown produce. It has been suggested that organic fruits and vegetables may carry higher levels of pathogenic bacteria since the fields are often fertilized with animal manure. However, fields in which crops are grown using conventional production practices may also be fertilized with manure. In a recent study, fruit and vegetable samples were collected at the farm level from 32 organic and 8 conventional farms (41). All samples were negative for *E. coli* O157:H7, whereas two samples from organic farms were positive for *Salmonella.* Average coliform counts were similar regardless of origin, conventional or organic. The prevalence of *E. coli* was six times greater on organic produce than on conventional produce. The increase in demand for bagged salad and organically grown produce provided the impetus for a study comparing the microflora of organically and conventionally grown spring mixes (which may contain radicchio, beet tops, baby red romaine, and arugula) (44). There were no statistical differences in populations of yeast, molds, coliforms, and psychrotrophic bacteria associated with conventional and organically grown spring mixes. Not surprisingly, populations of all microbes tested were higher on unwashed than on washed spring mixes. *E. coli* was associated with conventional and organically grown products; *Salmonella* and *L. monocytogenes* were not detected in any of the samples.

Preventing the contamination of raw fruits and vegetables preharvest is effectively impossible. Guidance documents to minimize the microbial contamination of fresh fruits and vegetables have been developed and have likely had an impact, if only by standardizing prevention measures (3). Recommendations for limiting the contamination of crops in the field include but are not limited to (i) microbiological testing of farm and irrigation water, (ii) construction of fencing to limit access of feral animals to fields and orchards, (iii) application of manure fertilizer to fields well in advance (>120 days) of planting, (iv) regular cleaning of farm equipment, and (v) use of surface rather than spray irrigation. Basically, these measures come under the evolving application of "good agricultural practices" (GAPs). Indeed, the driving force behind the development of GAPs is the microbiological safety of food (55). A recent study showed that the adoption of GAPs for reducing microbial contamination of crops was high among New England farmers; however, improvement needed to be made in record keeping and in washing and sanitizing of containers and food contact surfaces (11). A survey of New England consumers indicated that they would be willing to pay up to a $1.00 premium for a produce basket from a GAP-certified farm (45).

The microbiological safety of seed sprouts represents a unique challenge since contaminated seed seems to be a source of pathogens, including *Salmonella* and *E. coli* O157:H7. The conditions used for sprouting seeds are near ideal conditions for the growth of bacteria, including those associated with human illness. Sprouts were linked to a number of large outbreaks of foodborne illness in the late 1990s. The largest outbreak of *E. coli* O157:H7 infection ever recorded occurred in Japan and was linked to consumption of contaminated sprouts (57). The numerous outbreaks prompted the Centers for Disease Control and Prevention and the FDA to meet with sprout industry representatives to discuss this emerging food safety issue. Inclusion of antimicrobial compounds in growth medium and water used in the sprouting process was not effective in eliminating pathogens from sprouts (26). The recommendation was put forth that seeds be treated with 20,000 ppm of calcium hypochlorite to kill pathogens. However, even this treatment was found to be ineffective (25). The FDA developed regulatory guidelines for the sprout industry in 1999 based on the best scientific information available at the time (20). Seed sprouts are now considered a potentially hazardous food, a situation that further underscores the human health risks associated with the commodity. Chapter 6 provides an in-depth discussion of seed sprout safety.

Harvest

During harvest, crops can become contaminated with pathogens associated with harvesting equipment or farm workers. Workers' harvesting knives, if not cleaned and sanitized regularly, can spread pathogens from a single contaminated commodity to an entire container. Similarly, automated equipment can contaminate potentially all crops harvested from a given field. Automated equipment can also spread pathogens from field to field if the equipment is not cleaned and sanitized before moving from one field to the next. In general, regardless of the type of produce being harvested, from a microbiological safety standpoint the hazards are common.

In the United States, the majority (approximately 90%) of farms that grow fruits and vegetables harvest by hand (61). Therefore, field worker hygiene is integral to ensuring the microbiological safety of fresh fruits and vegetables. The degree of human handling varies by commodity and in some instances is increased as demand for a fresh product moves processing into the field. Maintaining worker hygiene in the field is difficult, but innovative measures including portable hand-washing stations and lavatory facilities have been developed.

The harvesting equipment, such as knives, clippers, and containers including trailers, boxes, bins, and truck beds, all contributes to the microflora of fruits and vegetables during harvest. Studies indicate that this equipment is washed prior to use about 75% of the time (61). Washing may remove gross debris but likely has little impact on reducing microbial loads. Sanitizing of equipment occurs even less frequently, about 30% of the time. Under this type of washing and sanitizing regime, bacteria, including pathogens, can build up on equipment, becoming extremely difficult to remove, and serve as a mechanism of contamination of the product. Reinforcing to field workers the importance of a regular schedule of washing and sanitizing of equipment will ensure that the practices become routine rather than an afterthought. The equipment being used is changing from harvest knives made of steel blades and wood handles to knives with plastic handles and stainless steel blades. Cleanable plastic and fiberglass bulk bins and containers are replacing those made of wood that are difficult to maintain in a clean and sanitized state (43).

Processing (Packing and Distribution)

Prior to reaching the consumer, most fruits and vegetables are washed at least once and usually several times during processing. In packing facilities, approximately 60% of the time waters used are treated with a sanitizer (61). Addition of sanitizer to processing waters is done to control the microbial load of the water rather than reduce the microbial load on the product. Indeed, water present as ice and that used for conveyance and washing can transfer microorganisms to the product. Water contacting produce should be potable since the products undergo minimal additional processing prior to consumption.

Food contact equipment should be washed and sanitized regularly. Unfortunately, research shows that tables, washing and cooling bins, and conveyor belts are washed and sanitized only about 50% of the time (61). Equipment entering a processing facility from the farm may have field soil containing human pathogens, on both the exterior and the interior. Potentially, the finished-product processing area could become contaminated with these pathogens. Sanitation is particularly important in plants where produce is processed as "fresh cut." A recent study monitoring the microbial load of lettuce during processing in an industrial plant showed that the shredding, rinsing, and centrifugation steps increased bacterial counts the most on the product (2). However, all steps influenced the microbial load.

Fresh-cut produce, such as prepackaged salads, has become a popular item at grocery stores. Consumers expect that no further washing of these products is required before serving these items. This expectation creates a significant challenge to the manufacturer to ensure the safety of such products, especially considering the volume of product sold. Sales of fresh-cut produce have grown dramatically from near zero in 1985 to $3.3 billion in 1994 and to $12.5 billion in 2004. Sales in 2005 are expected to reach $15 billion (6, 12). As technology advances, the sales of fresh-cut fruit are expected to reach $2 billion by 2008. New packaging materials and technologies will increase safety and shelf lives of these products. Indeed, the shelf life for fresh-cut vegetables held near 0°C is 15 to 21 days, depending on the commodity. Ultimately, pathogens must be controlled in the food-processing facility. This requires good manufacturing practices, a sanitation program, and development and implementation of a sound hazard analysis critical control point (HACCP) plan. Companies specializing in fresh-cut produce recognize that in order to succeed they must be vigilant about food safety. Part of minimizing food safety risks involves a hands-off handling approach and rapid cooling of the product to 4°C, after which the temperature should be maintained until the product reaches the customer (24). Presently, in the United States there are no industry standards for microbial populations on bagged lettuce. However, maintaining a low microbial load is advantageous in minimizing spoilage (prolonging shelf life) and promoting human health.

Food Preparation

Food preparation, whether in the home or at a restaurant (formal or quick service) or other type of food service facility (school cafeteria, nursing home, or hospital), can introduce pathogens into a product if not done properly. Salads and other produce are generally consumed raw, receiving no treatment prior to consumption that would kill pathogens associated with the product. Cases of foodborne illness linked to consumption of produce contaminated by food service employees are well documented (see "Outbreaks"). Poor employee hygiene practices, particularly the lack of hand washing after use of toilet facilities, can result in the spread of viruses and bacterial pathogens such as *Shigella*.

Each year, millions of cases of foodborne illness go unreported (51). Many of these cases are likely the result of improper preparation of foods in the home. *E. coli* O157:H7, *Salmonella, Campylobacter*, and *Vibrio* associated with raw meat, poultry, and seafood can cross-contaminate produce during preparation of salads and sandwiches. Knives used for cutting raw meats and seafood must be adequately cleaned before being used to cut lettuce, tomatoes, or other vegetables intended to be consumed raw. Separate cutting boards, one for meats and seafood and the other for fruits and vegetables, should ideally be used. Salmonellosis linked to contaminated cantaloupes prompted the recommendation that the outer rind of the cantaloupe be thoroughly washed, dried, and cut with a clean knife and gloved hands to prevent the contamination of the edible portion. Any portion of the cantaloupe not intended for immediate consumption should be wrapped and placed in a refrigerator.

Foodborne illness associated with improper handling of fruits and vegetables can be prevented through consumer education programs. Basic knowledge of proper handling, storage, and preparation should minimize microbial hazards associated with fresh fruits and vegetables and, for that matter, all foods prepared in the home. Many consumers do not consider that handling product in the home may result in pathogenic contamination (45). Food handling safety tips are often found in the produce section of grocery stores, on packages, and in newspapers and popular-press magazines.

The microbiological quality of ready-to-eat salads and prepared salads at retail facilities is in part a function of the hygiene practices of employees. Poor handling practices immediately prior to consumption can negate GAPs, good manufacturing practices, and the best HACCP programs implemented to ensure the microbiological safety of a product. Microbiological examination of nearly 3,000 samples of salad vegetables from food service areas and self-service salad bars revealed that 97% were of acceptable microbiological quality (49). *Salmonella, E. coli* O157, and *Campylobacter* were not detected in any of the samples examined; one sample was positive for *L. monocytogenes.* Therefore, most samples that were considered microbiologically unsatisfactory were so categorized because *E. coli* levels were in the range of 10^2 to 10^5 CFU/g.

Imported Products

Access to fresh produce year-round is no longer considered a luxury but is now seen as a necessity by food service and retail buyers alike. Perhaps not surprisingly, the United States is the world's largest importer of fresh fruits and vegetables. Products including lettuce, cantaloupes, stone fruits, raspberries, grapes, and avocados are shipped from the Southern Hemisphere to meet consumer demand during the Northern Hemisphere's winter. Fresh fruit imports increased by 155% and fresh vegetable imports increased by 265% between 1980 and 2001 (10). In 2001, the United States imported approximately $11 billion worth of fresh fruits and vegetables (12). Presently, imports account for approximately 27% of the volume of produce sold by grocery stores, and the proportion is expected to increase to one-third in 3 years (15). The increased consumption of ethnic foods will also drive the market for more expensive specialty fruits and vegetables. The importation of fresh fruits and vegetables creates unique food safety concerns.

The key issue among retailers with respect to imported fresh fruits and vegetables is ensuring food safety (15). Given the apparent increase in cases of foodborne illness linked to consumption of contaminated fresh fruits and vegetables, it is not surprising that microbial safety tops the list of retailers' priorities. The FDA, as mentioned previously, conducted a survey of imported fresh produce, and although the incidence of contamination was relatively low (approximately 4%) considering the shear volume of fresh fruits and vegetables imported each year, this level of contamination could pose a significant human health threat (http://www.cfsan.fda.gov/~dms/prodsur6.html; 4). The presence of *Salmonella* on tested produce may have been the result of workers' handling of the product, animal manures used as fertilizers, or harvesting equipment. A few samples were positive for *Shigella*, a human pathogen that is transmitted via the fecal-oral route, suggesting unsanitary handling by infected food workers. The results of the study underscored the need to better scrutinize farming practices in countries from which the contaminated product was imported.

The microbiological quality of fresh fruits and vegetables at retail markets is perhaps of greatest importance. In a study conducted by the FDA, total coliform counts, aerobic-plate counts, *E. coli* counts, yeast and mold counts, and the presence of human pathogens were investigated (58). Origin of the samples, domestic or imported, was not known. Samples were not positive for *Salmonella* or *Campylobacter. Staphylococcus aureus, E. coli,* and *Bacillus* were detected in single samples. *L. monocytogenes* was found in 6 (32%) of 19 samples tested for the pathogen. Pathogens such as *Campylobacter jejuni* survive quite well on fresh produce (33). Hirotani et al. (29) determined the presence of indicators of fecal contamination in vegetables sold in the United States and Mexico. The origin of the products purchased in the United States could not be determined. All of the samples purchased in Mexico and the United States tested positive for each of the fecal indicators: coliphages, fecal streptococci, total coliforms, and fecal coliforms. However, levels of fecal indicators were lower on United States-purchased products than on products purchased in Mexico.

OUTBREAKS

The FDA and the World Health Organization and Food and Agriculture Organization encourage the consumption of fresh fruits and vegetables as part of a healthy diet. The globalization of the food supply has introduced many regional foods to entirely new marketplaces. The consumption of fresh fruits and vegetables has increased consistently during the past 30 years. Unfortunately, the number of cases of foodborne illness linked to consumption of fresh fruits and vegetables has also increased (21). Outbreaks are not associated with one particular commodity or a specific microbe. In fact, within the past decade pathogens not typically associated with produce-related foodborne illness, including *E. coli* O157:H7 and *Cyclospora cayetanensis*, have caused large outbreaks.

In the United States during the period from 1973 to 1997, produce-associated outbreaks accounted for a growing proportion of all foodborne outbreaks, increasing from 0.7% of all outbreaks in the early 1970s to 6% in the 1990s (51, 61). Salads, sprouts, melons, lettuce, and berries were the most frequently implicated produce items, and those items remain at the top of the list of produce-associated outbreaks. In fact, 5.5% of foodborne outbreaks in England and Wales during the period from 1992 to 2000 were associated with the consumption of salad vegetables or fruit, similar to the value reported in the United States during the 1990s (39). Nearly 60% of illnesses were associated with *Salmonella* and norovirus. There are a number of review articles covering produce-related outbreaks in the United States, Canada, England, and other countries; therefore, this discussion will highlight trends and focus on the most recent outbreaks (39, 50, 51).

The popular press has focused particular attention on outbreaks associated with lettuce and tomatoes, prompting consumers to be particularly concerned about the safety of those foods. In an effort to address safety concerns, the FDA recommended that firms that grow, pack, or ship fresh lettuce and fresh tomatoes take action to enhance the safety of those products (22). The recommendation stemmed from the fact that 14 outbreaks accounting for approximately 859 cases of illness occurred from 1996 to 2003 (23). The causative agents associated with lettuce included *E. coli* O157:H7, *Cyclospora*, and hepatitis A virus but only *Salmonella* for tomatoes. The implicated products were of U.S. and non-U.S. origins.

The microbiological safety of produce is a global concern. A notable outbreak was attributed to the consumption of *E. coli* O157:H7-contaminated radish sprouts, in which a total of >10,000 people became ill. From 1989 to 1990, cantaloupe contaminated with *Salmonella enterica* serovar Chester was estimated to have affected 25,000 people in the United States. *Shigella sonnei* outbreaks in northern Europe and Denmark, in 1994 and 1998, respectively, were linked to contaminated lettuce from Spain and contaminated baby maize from Thailand. An outbreak of produce-associated botulism occurred in the United States in 1987 following the consumption of coleslaw (39). Most of the large outbreaks have been associated with bacterial pathogens; however, numerous cases of foodborne illness in the United States and Canada have been associated with *Cyclospora.* In Finland, imported salad items were linked to outbreaks of hepatitis A in 1996 and 1997; in 1998, raspberries contaminated with calicivirus were responsible for an outbreak involving hundreds of people. International outbreaks underscore how contamination of fresh fruits and vegetables during production or processing can have serious public health consequences over a large geographic area.

Bacteria

The potential for large outbreaks of *Salmonella* infection linked to consumption of contaminated tomatoes is a concern, considering that 5 billion pounds of fresh tomatoes are consumed annually in the United States. During the period from 1990 to 2004, the number of tomato-associated *Salmonella* outbreaks increased in frequency and magnitude (1,616 illnesses; nine outbreaks). In 2004, three outbreaks of *Salmonella* infection associated with eating roma tomatoes were detected in the United States and Canada (13). In total, 561 outbreak-related illnesses occurred in 18 states and one province in Canada. Collectively, for the three outbreaks, a number of *Salmonella* serotypes were identified: Javiana, Typhimurium, Anatum, Thompson, Muenchen, and Braenderup. No clear source of contamination was identified following environmental investigation of four packers and associated farms. Intervention strategies are inadequate to ensure that tomatoes are not contaminated with *Salmonella* or some other foodborne pathogen since knowledge of mechanisms of tomato contamination and methods of eradication of pathogens on produce is limited.

Produce is a prominent food vehicle of *E. coli* O157:H7 outbreaks in the United States (47). The association of *E. coli* O157:H7 with produce-associated outbreaks is a classic example of a pathogen emerging as a significant human health concern in a new food vehicle since outbreaks of infection with this pathogen were initially associated only with meat and meat products. The increase in association of *E. coli* O157:H7 with produce-related outbreaks is clearly demonstrated in Fig. 1. During the period from 1982 to 2002, produce was associated with 21% (38 of 183) of foodborne outbreaks and 34% of 5,269 cases of foodborne *E. coli* O157:H7 illness. Approximately 18% of produce-related outbreaks were linked to lettuce, and the outbreaks were not associated with kitchen-level cross-contamination. All outbreaks were associated with domestic produce (47).

Produce-related outbreaks are frequently associated with restaurants and other food service establishments. Restaurants accounted for the most *E. coli*

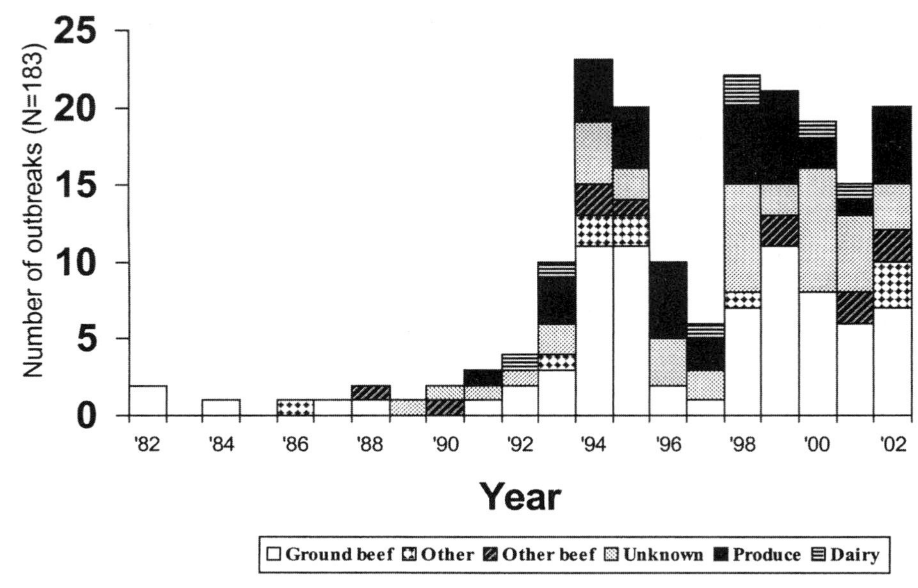

Figure 1 Food vehicles associated with *E. coli* O157:H7 outbreaks. (Reprinted from J. M. Rangel, P. H. Sparling, C. Crowe, P. M. Griffin, and D. L. Swerdlow, Epidemiology of *Escherichia coli* O157:H7 outbreaks, United States, 1982–2002, *Emerg. Infect. Dis.* **11**:603–609, 2005.)

O157:H7 produce-related associated outbreaks (40%; 15 of 38), and approximately 47% were the result of cross-contamination during food preparation (47). Contaminated carrots in salads served on several airline flights caused at least 45 cases of foodborne illness likely due to *Shigella*. The FDA cited the airline food supplier for filth following the discovery of live cockroaches, dirty utensils, dirty uncovered trash cans, and oozing pink slime in the food preparation area (http://www.fda.gov/foi/warning_letters/g5318d.pdf). Individually or collectively, such conditions may contribute to contamination of the product; however, the investigation did not determine when, where, or how the carrots became contaminated. Produce-linked cases of foodborne illness are also associated with improper handling practices in the home. Documenting consumer practices for handling food at home is difficult, and limited information is available (see chapter 7).

Outbreaks of produce-linked foodborne illness in the United States and in other countries are averted through routine testing of products by federal and state agencies. Red pear tomatoes were recalled following routine sampling by the FDA that revealed that the product may have been contaminated with *Salmonella* (http://www.fda.gov/oc/po/firmrecalls/caspecialty05_05.html). A recall of ground basil was initiated by a California company when routine testing of the product by the California Department of Health Services revealed the presence of *Salmonella* (http://www.fda.gov/oc/po/firmrecalls/american05_05.html). In both of the examples, no illnesses were reported in connection with products that may have reached consumers prior to the recalls.

Viruses

Viruses are a frequent cause of foodborne outbreaks, and many outbreaks of viral infection are attributed to consumption of contaminated fresh fruits and vegetables. In the United States from 1998 to 2000, 17% (13 of 76) of norovirus (previously known as "Norwalk-like virus")-associated outbreaks were linked to fruits and vegetables (63). Food handlers are often (48 of 94 outbreaks) implicated in outbreaks linked to viruses as opposed to outbreaks (20 of 102 outbreaks) involving bacteria. The importance of foodborne transmission of viruses varies widely by country; however, food was consistently identified as an important vehicle of transmission of viruses (39, 40).

In the United States in the past 5 years, three outbreaks associated with consumption of hepatitis A virus-contaminated raw green onions (scallions) resulted in approximately 800 cases of illness. The most recent outbreak involving green onions occurred in 2003, sickened >700 people, and resulted in three deaths. Contamination of the green onions was not associated with a food service worker but occurred at some point during distribution, growth, harvest, packing, or cooling (16). Crops that require extensive handling during harvest and processing for distribution to the consumer and that receive no additional treatment prior to being served raw or partially cooked present the greatest risk (19). This underscores the need for sanitary facilities (hand-washing stations and toilet facilities with potable water) for field workers. Development of new methods for detection of viruses in fruits and vegetables should aid in epidemiological investigations attempting to find the source or point of contamination and perhaps prevent contaminated products from reaching the consumer (17).

Recent reviews cover survival of viruses on crops in the field and on fruits and vegetables from harvest to consumption (35, 48). Viruses can remain infectious on crops for >30 days following irrigation with sludge or effluent. As with other microbes, irrigation water and manure fertilizer can serve as vehicles for dispersion of viruses onto crops in the field. Viruses inoculated onto vegetables, lettuce, carrots, radishes, cucumbers, and tomatoes could be recovered for up to 8 days postchallenge. Humidity plays an important role in survival of viruses on fresh fruits and vegetables. Under humid conditions and refrigeration (4°C), virus populations did not decrease on cherries or peaches, but under dry storage conditions, <1% of the originally seeded population could be recovered. Hepatitis A virus survived for 4, 7, and 9 days postchallenge on carrots, fennel, and lettuce, respectively (14). Enhanced persistence on lettuce was speculatively due to the wrinkled surface of the leaves, which provided greater protection for the virus. The authors of the report state that regardless of the commodity, washing does not substantially reduce viral contamination.

Protozoa

Foodborne outbreaks of cyclosporiasis and cryptosporidiosis have been associated with the consumption of fresh snow peas, raspberries, basil, and mesclun lettuce and a variety of other fresh fruits and vegetables (1, 7, 27, 31). The outbreaks that have occurred in the United States in recent years have been associated with consumption of contaminated fresh produce originating from various countries. The modes of contamination of the fresh produce have not been determined but may include use of contaminated water for irrigation, application of pesticides, and cleaning of product immediately following harvest (7). Irrigation waters from the United States and Central America were positive for a number of human pathogenic parasites, suggesting that use of contaminated water on crops that are traditionally consumed raw may result in human illness (59). Screening of food and water samples for protozoa can be particularly difficult since the parasites do not grow in the food and enrichment to increase numbers before detection is not possible. However, sensitive methods for detection of *Cryptosporidium* and *Cyclospora* associated with vegetables have been developed (34, 53).

Recognition in the United States of *Cyclospora* as a pathogen associated with fresh fruits and vegetables was precipitated by a large outbreak linked to consumption of contaminated imported raspberries (27). Outbreaks of cyclosporiasis linked to consumption of contaminated Guatemalan raspberries continue to occur. In 2000, 54 attendees at a wedding became ill with cyclosporiasis linked to consumption of wedding cake that had a cream filling that included raspberries (30). Officials from the FDA inspected the farm in Guatemala where the raspberries were grown and found no definitive source of contamination. During the summer of 2004, approximately 50 cases of cyclosporiasis were associated with consumption of contaminated snow peas (7). The raw snow peas were washed in municipal water and added to a salad. The snow peas were imported from Guatemala; the route of contamination was not determined. Prior to this outbreak, snow peas had not been implicated as a vehicle of *Cyclospora.* Data from the U.S. Foodborne Diseases Active Surveillance Network (FoodNet) of the Centers for Disease Control and Prevention's Emerging Infections Program suggest that the incidence of infections caused by *Cryptosporidium* declined in 2004 and that few cases ($n = 15$) of cyclosporiasis occurred (8). However, outbreaks of both cryptosporidiosis and cyclosporiasis continue to occur and efforts must progress to further decrease numbers of illnesses linked to these human pathogens. In early 2005, fresh basil was implicated as the source of illness in 293 laboratory-confirmed cases of cyclosporiasis in Florida. Outbreaks of cryptosporidiosis and cyclosporiasis in the United States are generally associated with imported fresh fruits and vegetables, making international collaboration critical to the successful identification of appropriate prevention and control measures.

THE FUTURE—WHAT'S NEXT?

There exist a plethora of microorganisms that may potentially contaminate fresh fruits and vegetables; however, only a small percentage of those are pathogenic to humans. There is no definitive means to predict what pathogen may emerge as a significant human health concern in association with the consumption of fresh fruits and vegetables. Indeed, 15 years ago there was little or no concern that *E. coli* O157:H7 would be linked to human illness resulting from the consumption of contaminated produce. In the increasingly global marketplace, agricultural and processing practices linked to fresh fruits and vegetables must be examined to develop methods to minimize the import or export of produce that may be contaminated with human pathogens. This is particularly important for fresh fruits and vegetables since they are generally consumed raw.

Government agencies, academia, and industry are investigating and developing new methods to enhance the microbiological safety of produce. Implementation of GAPs and HACCP programs is an important measure for reducing microbial loads on produce. International cooperation will facilitate the utilization of the best production practices and aid in the investigation of outbreaks in an effort to identify effective control measures. Many new technologies are being developed for use postharvest to reduce microbial numbers on produce (see chapters 4 and 5). Consumer awareness of proper storage and handling practices of fresh fruits and vegetables must be enhanced through education programs since the consumer is ultimately the last line of defense. A study of New England consumers found that few thought in-home handling practices were a significant source of pathogenic contamination (45). Measures to ensure the microbiological safety of fresh fruits and vegetables have advanced significantly in the past decade; consumption of fruits and vegetables must continue to be encouraged as part of a healthy diet.

REFERENCES

1. **Alakpa, G. E., S. C. Clarke, and A. F. Fagbenro-Beyioku.** 2003. *Cyclospora cayetanensis* infection: vegetables and water as possible vehicles for its transmission in Lagos, Nigeria. *Br. J. Biomed. Sci.* **60:**113–114.
2. **Allende, A., E. Aguaya, and F. Artes.** 2004. Microbial and sensory quality of commercial fresh processed red lettuce throughout the production chain and shelf life. *Int. J. Food Microbiol.* **91:**109–117.
3. **Anonymous.** 1998. *Guide To Minimize Microbial Food Safety Hazards for Fresh Fruits and Vegetables.* Center for Food Safety and Applied Nutrition, Food and Drug Administration, U.S. Department of Health and Human Services, Washington, D.C.
4. **Beru, N., and P. A. Salbury.** 2002. FDA's produce safety activities. *Food Safety Magazine* **2002**(Feb.–Mar.): 14–19.
5. **Beuchat, L. R.** 2002. Ecological factors influencing survival and growth of human pathogens on raw fruits and vegetables. *Microbes Infect.* **4:**413–423.
6. **Brody, A. L.** 2005. What's fresh about fresh-cut. *Food Technology* **59:**74–77.
7. **Centers for Disease Control and Prevention.** 2004. Outbreak of cyclosporiasis associated with snow peas—Pennsylvania, 2004. *Morb. Mortal. Wkly. Rep.* **53:**876–878.
8. **Centers for Disease Control and Prevention.** 2005. Preliminary FoodNet data on the incidence of infection with pathogens transmitted commonly through food—10 sites, United States, 2004. *Morb. Mortal. Wkly. Rep.* **54:**352–356.
9. **Chalmers, R. M., H. Aird, and F. J. Bolton.** 2000. Waterborne *Escherichia coli* O157. *J. Appl. Microbiol.* **88:**124S–132S.
10. **Clemens, R.** 2004. The expanding U.S. market for fresh produce. *Iowa Agric. Rev.* **10:**1–4.
11. **Cohen, N., C. S. Hollingsworth, R. Brennan Olson, M. J. Laus, and W. M. Coli.** 2005. Farm food safety practices; a survey of New England growers. *Food Prot. Trends* **25:**363–370.
12. **Cook, R. L.** 2003. The evolving global marketplace for fruits and vegetables. *Calif. Agric.* **2003**(Jan. –Mar.):1–3.
13. **Cordy, R., V. Lanni, V. Kistler, V. Dato, A. Weltman, C. Yozviak, K. Waller, K. Nalluswami, M. Moll, J. Lockett, M. Lynch, C. Braden, S. K. Gupta, and A. DuBois.** 2005. Outbreaks of *Salmonella* infections associated with eating Roma tomatoes—United States and Canada, 2004. *Morb. Mortal. Wkly. Rep.* **54:**325–328.
14. **Croci, L., D. De Medici, C. Scalfaro, A. Fiore, and L. Toti.** 2002. The survival of hepatitis A virus in fresh produce. *Int. J. Food Microbiol.* **73:**29–34.
15. **Cuellar, S.** 2003. Marketing fresh fruit and vegetable imports in the United States: status, challenges and opportunities. *SMART Marketing.* **2003**(Mar.):6–11.
16. **Dato, V., A. Weltman, K. Waller, M. A. Ruta, C. Hembree, S. Evenson, C. Wheeler, and T. Vogt.** 2003. Hepatitis A outbreak with green onions at a restaurant—Monaca, Pennsylvania, 2003. *Morb. Mortal. Wkly. Rep.* **52:**1155–1157.

17. **Dubois, E., C. Agier, O. Traore, C. Hennechart, G. Merle, C. Cruciere, and H. Laveran.** 2002. Modified concentration method for the detection of enteric viruses on fruits and vegetables by reverse transcriptase-polymerase chain reaction or cell culture. *J. Food Prot.* **65**:1962–1969.
18. **Findlar, C. R.** 1971. The survival of *Salmonella* Dublin in cattle slurry. *Vet. Rec.* **89**:224–225.
19. **Fiore, A. E.** 2004. Hepatitis A transmitted by food. *Clin. Infect. Dis.* **38**:705–715.
20. **Food and Drug Administration.** 1999. Guidance for industry: reducing microbial food safety hazards for sprouted seeds and guidance for industry: sampling and microbial testing of spent irrigation water during sprout production. *Fed. Regist.* **64**:57893–57902.
21. **Food and Drug Administration and Center for Food Safety and Applied Nutrition.** 30 September 2001, posting date. *Analysis and Evaluation of Preventative Control Measures for the Control and Reduction/Elimination of Microbial Hazards on Fresh and Fresh-Cut Produce,* chapter 4. [Online.] Food and Drug Administration and Center for Food Safety and Applied Nutrition, Washington, D.C. http://www.cfsan.fda.gov/~comm/ift3-4a.html.
22. **Food and Drug Administration and Center for Food Safety and Applied Nutrition.** 5 February 2004, posting date. *Letter to Firms That Grow, Pack, or Ship Fresh Lettuce and Fresh Tomatoes.* [Online.] Food and Drug Administration and Center for Food Safety and Applied Nutrition, Washington, D.C. http://www.foodsafety.gov/~dms/prodltr.html.
23. **Food and Drug Administration and Center for Food Safety and Applied Nutrition.** 18 June 2004, posting date. *Produce Safety from Production to Consumption: a Proposed Action Plan To Minimize Foodborne Illness Associated with Fresh Produce Consumption.* [Online.] Food and Drug Administration and Center for Food Safety and Applied Nutrition, Washington, D.C. http://www.foodsafety.gov/~dms/prodplan.html.
24. **Gale, S. F.** 2004. Fresh express: cutting-edge food safety. *Food Safety Magazine* **2004** (February–March):51–54.
25. **Gandhi, M., and K. R. Matthews.** 2003. Efficacy of chlorine and calcinated calcium treatment of alfalfa seeds and sprouts to eliminate *Salmonella. Int. J. Food Microbiol.* **87**:301–306.
26. **Gandhi, M., S. Golding, S. Yaron, and K. R. Matthews.** 2001. Use of green fluorescent protein expressing *Salmonella* Stanley to investigate survival, spatial location, and control on alfalfa sprouts. *J. Food Prot.* **64**:1891–1898.
27. **Herwaldt, B. L., M. J. Beach, and Cyclospora Working Group.** 1997. An outbreak in 1996 of cyclosporiasis associated with imported raspberries. *N. Engl. J. Med.* **336**:1548–1556.
28. **Himathongkham, S., S. Bahari, H. Riemann, and D. Cliver.** 1999. Survival of *Escherichia coli* O157:H7 and *Salmonella* Typhimurium in cow manure and cow manure slurry. *FEMS Microbiol. Lett.* **178**:251–257.
29. **Hirotani, H., J. Naranjo, P. G. Moroyqui, and C. P. Gerba.** 2002. Demonstration of indicator microorganisms on surface of vegetables on the market in the United States and Mexico. *J. Food Sci.* **67**:1847–1850.
30. **Ho, A. Y., A. S. Lopez, M. G. Eberhart, R. Levenson, B. S. Finkel, A. J. da Silva, J. M. Roberts, P. A. Orlandi, C. C. Johnson, and B. L. Herwaldt.** 2002. Outbreak of cyclosporiasis associated with imported raspberries, Philadelphia, Pennsylvania, 2000. *Emerg. Infect. Dis.* **8**:783–788.

31. **Hoang, L. M., M. Fyfe, C. Ong, J. Harb, S. Champagne, B. Dixon, and J. Isaac-Renton.** 2005. Outbreak of cyclosporiasis in British Columbia associated with imported Thai basil. *Epidemiol. Infect.* **133**:23–27.
32. **Institute of Food Technologists.** 20 February 2002, posting date. *IFT Expert Report on Emerging Microbiological Food Safety Issues: Implications for Control in the 21st Century.* [Online.] Institute of Food Technologists, Chicago, Ill. http://www.ift.org/pdfs/expert/microfs/webreport.pdf.
33. **Karenlampi, R., and M. L. Hannien.** 2004. Survival of *Campylobacter jejuni* on various fresh produce. *Int. J. Food Microbiol.* **97**:187–195.
34. **Kniel, K. E., and M. C. Jenkins.** 2005. Detection of *Cryptosporidium parvum* oocysts on fresh vegetables and herbs using antibodies specific for *Cryptosporidium parvum* viral antigen. *J. Food Prot.* **68**:1093–1096.
35. **Koopmans, M., C. H. von Bonsdorff, J. Vinje, D. de Medici, and S. Monroe.** 2002. Foodborne viruses. *FEMS Microbiol. Rev.* **26**:187–205.
36. **Kruse, H., A.-M. Kirkemo, and K. Handeland.** 2004. Wildlife as source of zoonotic infections. *Emerg. Infect. Dis.* **10**:2067–2072.
37. **Kudva, I., K. Blanch, and C. J. Hovde.** 1998. Analysis of *Escherichia coli* O157:H7 survival in ovine or bovine manure and manure slurry. *Appl. Environ. Microbiol.* **64**:3166–3174.
38. **Lin, B.-H.** 2004. Fruit and vegetable consumption: looking ahead to 2020. *Agric. Infect. Bull.* **2004**:792–797.
39. **Long, S. M., G. K. Adak, S. J. O'Brien, and I. A. Gillespie.** 2002. General outbreaks of infectious intestinal disease linked with salad vegetables and fruit, England and Wales, 1992–2000. *Communicable Dis. Public Health* **5**:101–105.
40. **Lopman, B. A., M. H. Reacher, Y. van Duijnhoven, F. Hanon, D. Brown, and M. Koopmans.** 2003. Viral gastroenteritis outbreaks in Europe, 1995–2000. *Emerg. Infect. Dis.* **9**:90–96.
41. **Mukherjee, A., D. Speh, E. Dyck, and F. Diez-Gonzalez.** 2004. Preharvest evaluation of coliforms, *Escherichia coli, Salmonella*, and *Escherichia coli* O157:H7 in organic and conventional produce grown by Minnesota farmers. *J. Food Prot.* **67**:894–900.
42. **Okafo, C. N., V. J. Umoh, and M. Galadima.** 2003. Occurrence of pathogens on vegetables harvested from soils irrigated with contaminated streams. *Sci. Total Environ.* **311**:49–56.
43. **Pabrua, F. F., and J. William.** 2003–2004. Challenges, progress and solutions in produce safety. *Food Safety Magazine* **2003–2004**(December–January):49–52.
44. **Phillips, C. A., and M. A. Harrison.** 2005. Comparison of the microflora on organically and conventionally grown spring mix from a California processor. *J. Food Prot.* **68**:1143–1146.
45. **Pivarnik, L. F., H. Donath, M. S. Patnoad, and C. Roheim.** 2005. New England consumers' willingness to pay for fresh fruits and vegetables grown on GAP-certified farms. *Food Prot. Trends* **25**:256–266.
46. **Plym-Forshell, L., and I. Ekesbo.** 1996. Survival of salmonellas in urine and dry faeces from cattle—an experimental study. *Acta Vet. Scand.* **37**:127–131.
47. **Rangel, J. M., P. H. Sparling, C. Crowe, P. M. Griffin, and D. L. Swerdlow.** 2005. Epidemiology of *Escherichia coli* O157:H7 outbreak, United States, 1982–2002. *Emerg. Infect. Dis.* **11**:603–609.

48. **Rzezutka, A., and N. Cook.** 2004. Survival of human enteric viruses in the environment and food. *FEMS Microbiol. Rev.* **28**:441–453.

49. **Sagoo, S. K., C. L. Little, and R. T. Mitchell.** 2003. Microbiological quality of open ready-to-eat salad vegetables: effectiveness of food hygiene training of management. *J. Food Prot.* **66**:1581–1586.

50. **Sewell, A. M., and J. M. Farber.** 2001. Foodborne outbreaks in Canada linked to produce. *J. Food Prot.* **64**:1863–1877.

51. **Sivapalasingam, S., C. R. Friedman, L. Cohen, and R. V. Tauxe.** 2004. Fresh produce: a growing cause of outbreaks of foodborne illness in the United States, 1973 through 1997. *J. Food Prot.* **67**:2342–2353.

52. **Solomon, E. B., C. J. Potenski, and K. R. Matthews.** 2002. Effect of irrigation method on transmission to and persistence of *Escherichia coli* O157:H7 on lettuce. *J. Food Prot.* **65**:673–676.

53. **Steele, M., S. Unger, and J. Odumeru.** 2003. Sensitivity of PCR detection for *Cyclospora cayetanensis* in raspberries, basil, and mesclun lettuce. *J. Microbiol. Methods* **54**:277–280.

54. **Steele, M., and J. Odumeru.** 2004. Irrigation water as source of foodborne pathogens on fruit and vegetables. *J. Food Prot.* **67**:2839–2849.

55. **Stier, R. F., and N. E. Nagle.** 2001. Growers beware: adopt GAPs or else. *Food Safety Magazine* **2001**(October–November):26–32.

56. **Tannock, G. W., and J. M. Smith.** 1972. Studies on the survival of *Salmonella* Typhimurium and *Salmonella* bovis-morbificans on soil and sheep faeces. *Res. Vet. Sci.* **13**:150–153.

57. **Taormina, P. J., L. R. Beuchat, and L. Slutsker.** 1999. Infections associated with eating seed sprouts: an international concern. *Emerg. Infect. Dis.* **5**:626–634.

58. **Thunberg, R. L., T. T. Tran, R. W. Bennett, R. N. Matthews, and N. Belay.** 2002. Microbial evaluation of selected fresh produce obtained at retail markets. *J. Food Prot.* **65**:677–682.

59. **Thurston-Enriquez, J. A., P. Watt, S. E. Dowd, R. Enriquez, I. L. Pepper, and C. P. Gerba.** 2002. Detection of protozoan parasites and microsporidia in irrigation waters used for crop production. *J. Food Prot.* **65**:378–382.

60. **Tournas, V. H.** 2005. Spoilage of vegetable crops by bacteria and fungi and related health hazards. *Crit. Rev. Microbiol.* **31**:33–44.

61. **U.S. Department of Agriculture-National Agricultural Statistics Service.** 13 June 2001, posting date. *Fruit and Vegetable Agricultural Practices—1999, June.* [Online.] U.S. Department of Agriculture-National Agricultural Statistics Service, Washington, D.C. http://usda.mannlib.cornell.edu/reports/nassr/other/pcu-bb/agfv0601.pdf.

62. **Wang, G., and M. P. Doyle.** 1998. Survival of enterohemorrhagic *Escherichia coli* O157:H7 in water. *J. Food Prot.* **61**:662–667.

63. **Widdowson, M.-A., A. Sulka, S. N. Bulens, R. S. Beard, S. S. Chaves, R. Hammond, E. D. P. Salehi, E. Swanson, J. Totaro, R. Woron, P. S. Mead, J. S. Bresee, S. S. Monroe, and R. I. Glass.** 2005. Norovirus and foodborne disease, United States, 1991–2000. *Emerg. Infect. Dis.* **11**:95–102.

Microbiology of Fresh Produce
Edited by Karl R. Matthews

Role of Good Agricultural Practices in Fruit and Vegetable Safety

2

Elizabeth A. Bihn and Robert B. Gravani

Keeping fruits and vegetables free from contamination by pathogenic organisms such as bacteria, viruses, and parasites, as well as from chemical and physical hazards, is important to the health and safety of every person who consumes produce (7, 42). Fruits and vegetables are at risk for contamination all throughout the production and distribution network, as well in retail stores, in restaurants, and in the home. The source of pathogens can be virtually anything in the production environment that comes in contact with produce (4). The primary source of most enteric pathogens on fresh produce is animal or human fecal material. Potential direct and indirect contamination can result from contact with the soil; manure; irrigation water; wild and domestic animals; farm, packinghouse, and terminal market workers; contaminated equipment in the field, packinghouse, and distribution system; wash, rinse, and flume water; ice; cooling equipment; and transportation vehicles or from cross contamination from other foods or improper storage, packaging, display, and preparation (2, 8, 9, 16, 32).

Current research indicates that some pathogens can now survive in low-pH environments, such as human stomach acid, where before these pathogens would have been killed. Pathogenic bacteria such as *Escherichia coli* O157:H7 can develop acid resistance while growing in the colons of grain-fed cattle (13). Manure from these animals may contain this acid-resistant pathogen and be used as a soil amendment for the production of fresh fruits and vegetables. Improper application time prior to harvest or improper composting may lead to contamination of these produce crops should they touch the soil.

Elizabeth A. Bihn and Robert B. Gravani, Department of Food Science, Cornell University, Ithaca, NY 14853.

Food selection and eating trends often provide new modes of transmission. It is estimated that 30 to 40% of all cancers can be prevented by diet and lifestyle alone (14). Consuming fruits and vegetables provides fiber and many nutrients that are essential parts of a healthy diet (31). With advances in agronomic practices, preservation technologies, and shipping practices; improved cold-chain management; and global production and distribution of fresh fruits and vegetables, many produce items are available year-round (31). The selection and diversity of fruits and vegetables are also increasing due to these advances. Today, consumers in a typical retail food store can select from an average of 345 different produce items that come from over 130 countries around the world (27, 32).

Many of these produce commodities are eaten raw, so they never receive heat treatment to kill pathogenic microorganisms that may be present. Although they provide many health benefits, raw fruits and vegetables also have been known to be potential vehicles for human disease (2). Fresh produce can serve as a source for all classes of foodborne pathogens, including bacteria, viruses, protozoa, fungi, and helminths, but pathogenic bacteria present the greatest concern because the risk that they pose may be amplified by growth prior to consumption (30). Finally, many consumers do not follow the U.S. Department of Agriculture recommendation to wash all fruits and vegetables with cool, clean water prior to consumption, so even this step meant to remove soil and particulate matter does not occur. An increase in fresh produce consumption, coupled with improper handling prior to consumption, provides an opportunity for the transmission of pathogenic microorganisms to people, resulting in the possibility of foodborne illnesses.

Host susceptibility is an additional issue in a population in which many individuals are immunocompromised due to a multitude of conditions. Many individuals are now living with human immunodeficiency virus infection or other chronic conditions or use immunosuppressive agents due to chemotherapy or organ transplantation. Simple aging as part of life's progression, pregnancy, youth, and residence in long-term care facilities can also lead to a compromised immune system (28, 43). When the numbers of people in these various situations are added together, it can be estimated that approximately 25% of the U.S. population is immunocompromised (43). These individuals are at an increased risk of infection from many sources, including fresh produce. Understanding relevant topics such as changes in the pathogens, food selection, and host susceptibility will assist in an objective discussion of produce food safety and implementation of "good agricultural practices" (GAPs). GAPs are defined as any operational procedures or activities that reduce microbial risks to fruits and vegetables on the farm or in the packinghouse. This chapter will focus on research findings and GAPs implementation that is being achieved on the farm and provide additional information to make GAPs implementation more effective and efficient.

RISKS ON THE FARM

Fruits and vegetables become contaminated with pathogenic microorganisms as well as unintended chemicals and physical hazards in many ways. Most produce commodities are produced in open fields and exposed to rain, soil, domestic and wild animals (including birds), and variable weather conditions. Besides naturally occurring elements, producing fruits and vegetables involves a complex system of additional inputs. The microbiological quality of water is always a concern because of the prevalence and importance of water in the production of fresh produce. Water is vital to plant growth, and many crops receive supplemental irrigation and protective topical sprays that are mixed with water. In addition, many commodities are cooled, moved, or washed with water prior to sale.

Contamination of produce with animal manure is another major concern since many animals shed pathogenic microorganisms in their manure. Some manure is applied as a soil amendment to promote soil health and fertility, but other manure, particularly from wild animals, is deposited in the field without the grower's knowledge and at various times during the production season. It is not only animal manure that is an issue but human feces as well. Many pathogenic microorganisms, such as hepatitis A virus and *Shigella* spp., have only human reservoirs, so when outbreaks caused by these organisms occur, they are almost certainly related to poor personal hygiene or improper waste disposal. Proper disposal of human and plant waste is part of a good sanitation program. Lack of a sanitation program or sporadic use of a sanitation program can lead to contamination of produce. Humans are involved in nearly every aspect of cultivated production. Areas where human hygiene and training are most important to produce food safety are those where there is direct contact with ripe, ready-to-eat produce during hand harvesting, sorting, grading, and packing.

Water quality, manure use, sanitation, and worker training and hygiene are important considerations in the production of fruits and vegetables since contamination has been linked to these factors (15, 17, 25, 26, 48). GAPs development can be focused here to minimize microbial risks.

Each farm operation is unique in terms of number of acres in production, commodities grown, production practices, and marketing outlets. Some commodities have been implicated in more foodborne illness outbreaks than others, and some have natural characteristics that make them more susceptible to contamination. With such variation in risk levels, in commodities, and in operations, it would be difficult to develop one set of standards that could be effective at reducing risks for all commodities and for every farm. A food safety program that is successfully applied at the farm level should take this variation into consideration and provide guidance to growers so that they can evaluate their own operations.

In 1998, the Food and Drug Administration (FDA) introduced the *Guide To Minimize Microbial Food Safety Hazards for Fresh Fruits and Vegetables* (16). The guide outlines a set of production practices termed GAPs and intervention strategies that can be applied on the farm to fresh produce production. GAPs are similar to "good manufacturing practices" (GMPs) used in the food processing industry but address agricultural activities including preplanting, planting, harvest, and postharvest practices designed to reduce microbial risks. The guide offers a broad range of guidelines that encompass many of the risk areas that growers would need to evaluate in order to implement a farm food safety program and urge growers, through proper management, to minimize potential contamination. The guide (1a, 16) provides the following eight principles of microbial food safety that can be applied to the growing, harvesting, packing, and transportation of fresh fruits and vegetables.

Principle 1: The prevention of microbial contamination of fresh produce is favored over reliance on corrective actions once contamination has occurred.

Principle 2: To minimize microbial food safety hazards for fresh produce, growers or packers should use GAPs in those areas over which they have a degree of control while not increasing other risks to the food supply or the environment.

Principle 3: Anything that comes in contact with fresh produce has the potential of contaminating it. For most foodborne pathogens associated with produce, the major source of contamination is human or animal feces.

Principle 4: Whenever water comes in contact with fresh produce, its source and quality dictate the potential for contamination.

Principle 5: Agricultural practices using manure or municipal biosolid wastes should be closely managed to minimize the potential for microbial contamination of fresh produce.

Principle 6: Worker hygiene and sanitation practices during production, harvesting, sorting, packing, and transportation play a critical role in minimizing the potential for microbial contamination of fresh produce.

Principle 7: Follow all applicable local, state, and federal laws and regulations, or corresponding or similar laws, regulations, or standards for agricultural practices for operators outside the United States.

Principle 8: Accountability at all levels of the agricultural environment (farms, packing facilities, distribution centers, and transport operations) is important to a successful food safety program. There must be qualified personnel and effective supervision to ensure that all elements of the program function correctly and to help track produce back through the distribution channels to the producer.

These principles are quite general given the broad range of fruits and vegetables and growing conditions. Like GMPs, they focus on minimizing microbial contamination (31). It should also be noted that the guide focuses only on microbial hazards for fresh produce and does not specifically address pesticide residues or chemical contaminants. The guide also focuses on risk reduction, not risk elimination, since present technologies cannot eliminate all potential food safety hazards associated with fresh produce that will be eaten raw (1a, 16).

Choosing To Implement GAPs on the Farm

Presently, GAPs are not regulations but were developed by the FDA to provide guidance and recommendations to fruit and vegetable growers. This is not to say that they will not become regulations if the industry does not work to reduce microbial risks in fruit and vegetable production. An example of the emergence of a new regulation is demonstrated by the development of the juice hazard analysis critical control point (HACCP) regulation. As a result of several highly publicized foodborne illness outbreaks resulting from consumption of unpasteurized cider in 1996, on January 19, 2001, the FDA published the "HACCP Procedures for the Safe and Sanitary Processing and Importing of Juice Final Rule" in the *Federal Register* (11a). This regulation stipulates that all fresh juice sold wholesale or imported must receive some type of treatment to ensure a 5-log reduction of the pertinent pathogen in finished juice prior to bottling. Producers that sell directly to consumers are exempt from this regulation, but due to several additional outbreaks linked to the consumption of unpasteurized or untreated cider, a number of organizations representing cider producers have recently recommended that all cider receive a treatment to reduce the presence of pathogenic organisms, thereby reducing the risk to consumers. Mandatory regulations are one consideration, but marketing also can be affected by outbreaks.

Media attention to foodborne illness outbreaks takes information directly to consumers. Even if outbreaks are occurring in states or regions far from their locality, consumers will avoid commodities that are involved in outbreaks. The 1996 *Cyclospora cayetanensis* outbreak involving raspberries resulted in an estimated loss of $50 million for the strawberry industry because it was mistakenly mentioned that strawberries were thought to be involved in the outbreak. A cider-associated outbreak in 1996 resulted in an estimated drop in share value of 41% for the company that produced the cider. A series of cantaloupe contaminations from imported Mexican cantaloupes caused the FDA to ban the import of all cantaloupes from Mexico until farms could independently prove to the FDA that they were utilizing practices that reduced microbial risks and satisfied other FDA concerns. Outbreaks directly affect the fresh produce business by influencing consumers, markets, sales, and overall profitability. As with other production practices that are done to increase quality and marketability of commodities, implementing GAPs and reducing microbial risks directly influence the marketability of fresh fruits and vegetables.

Unless their commodity has been directly affected by a foodborne illness outbreak, most growers are spurred to implementation because of buyer expectations. In 2000, several large retail buyers began sending letters to their produce suppliers requiring that the produce they sell be produced using GAPs. In addition to the use of GAPs to minimize microbial contamination of fresh fruits and vegetables, some produce buyers have introduced purchasing specifications, letters of guarantee, vendor certification programs, and the requirement for independent, third-party audits of farms and packinghouses to provide assurance that growers are following GAPs (23, 24, 31). For many growers, these new requests and requirements were unexpected and left many wondering how best to meet these new demands. The additional cost of an audit was also a concern, and depending on the auditing firm, the size of the farm, distribution of the farmland, and other factors, the cost of the audits could be quite substantial. Growers are forced to weigh the value of the contracts versus the cost of GAPs implementation.

To clearly illustrate the relationship of GAPs to produce safety, the produce safety assurance pyramid was developed (Fig. 1). There are four main components in the pyramid. At the top of the pyramid, produce safety depends on the commitment and leadership of top management in a company. Management needs to believe in the importance of produce safety and provide the necessary resources (finances, personnel, and equipment, etc.) to achieve this goal. A well-developed and effective education and training program throughout the company is vital to produce safety. Knowledgeable farm, packinghouse, and transportation workers who consistently perform appropriate tasks correctly will reduce microbial risks. A thorough understanding of biological hazards is a key factor in controlling them. Hazards that are not properly identified or understood cannot be easily addressed. The produce safety assurance pyramid is built on a strong foundation of GAPs, including practices that control potential hazards on the farm. In the

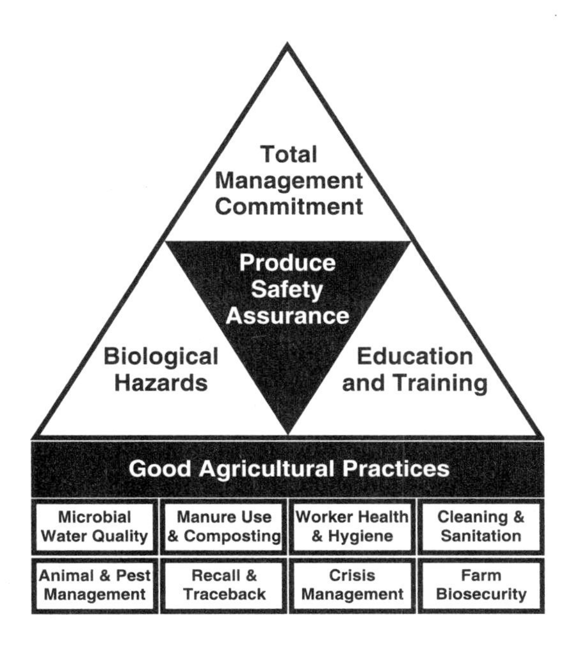

Figure 1 Produce safety assurance is built on a solid foundation of GAPs. (Figure used with permission from Robert B. Gravani.)

pyramid, the relevant areas are presented as microbial water quality, manure use and composting, worker health and hygiene, and cleaning and sanitation, as well as animal and pest management, recall and trace-back strategies, crisis management, and farm biosecurity. An important link between these GAPs components is an effective record-keeping system, which details the tasks that were performed.

GAPs Education and Training Programs

As with any food safety program, the first step to implementation occurs when growers and farm managers make GAPs a priority at the management level of the operation. Providing the resources and the vision for the food safety program will make implementation part of the operation and will set an example for all employees, showing them that the farm is committed to supporting all efforts to properly implement GAPs. The discussion of GAPs and food safety at the farm level is relatively new, and several challenges exist.

Many produce growers have been in the business for more than 20 years, and adapting to new growing practices can be a difficult transition. Many have the following response to this change: "I have been doing it this way for 20 years and have never made anyone sick. Why do I need to change now?" It is a reasonable question with several answers that often make teaching and training programs more effective. Many foodborne illnesses go unreported, particularly if the sick individual is normally in good health. Most healthy people will not go to a doctor or provide a sample to verify the presence of a foodborne pathogen. Many illnesses are hard to trace, particularly in produce because of the short shelf life of the product. In conveying the importance of produce food safety to fruit and vegetable growers, farm workers, and packers, it is extremely important to include information about changing demographics, changes in food choices, and changes in the microorganisms of concern in fruit and vegetable production. In fact, an increase in illnesses may not be related to changes in production but rather to changes in all of these other factors. It is important to remember that most of the GAPs education and training programs are targeted at adults, who learn better when specific information that is pertinent to the objectives is shared.

Adult learners enter learning situations with their own goals, motivations, needs, and experiences (19). They are motivated to learn when they know exactly what to expect from an education and training program and when the information is relevant to them. At the beginning of a program, adult learners need to know how they will personally benefit from the information presented. They also need to know *why* they are learning a new skill or are asked to perform a new task. Assuming that employees will "do what you say because you said so" ignores the principles of adult education. Adults are motivated when training activities that include opportunities for involvement and interaction are built into the program. They also appreciate being able to practice the information or skills that they have learned in team exercises, scenario case studies, and other activities (19).

Learning is affected by many factors, including the following:

- Capability and attitude of the learners
- Educational background, previous training, and experience of the learners
- Nature of the material being taught
- Instructional method(s)
- Capability and attitude of the instructor

Effective GAPs education and training programs can be successfully developed and implemented if trainers remember some basic principles of adult education, especially that adults learn better under the following conditions.

- They are involved.
- The materials are structured to meet their needs.
- The subject is connected to their daily activities.
- They work together in informal, organized groups.
- Materials are presented through a variety of methods with practical examples that explain expectations and procedures.

- They are given opportunities to apply and practice what they have learned.

For GAPs education and training programs with fruit and vegetable growers and farm or packinghouse workers, trainers must know and understand the following three main guidelines.

1. Know the subject. Trainers should decide what knowledge and skills the participants should possess after the program is completed. These decisions can be based on the present situation and any problems that may be occurring in the field or packinghouse.
2. Know the audience. It is vital to know as much as possible about the adult learners in the audience. Presenting training in the native language of the employees and understanding cultural norms that may or may not be appropriate on U.S. farms are important.
3. Discuss the program goals and objectives. Trainers should present the goals and objectives of the program and the reason why the information is important. The benefits to each participant should be explained clearly. The acronym WIIFM ("What's in it for me") is an important reminder to explain how the program benefits everyone. Many GAPs that are implemented provide direct benefits to farm workers by preventing illness and exposure to anything that may be on their hands due to field work.

Importance of Worker Health and Hygiene

Several foodborne illness outbreaks have been the result of workers' contaminating food prior to consumption (10, 11). Workers who contaminate produce are usually infected with a pathogen (such as hepatitis A virus, *E. coli* O157:H7, and *Shigella* spp.) and likely do not follow proper hygiene steps such as effective hand washing. Preventing contamination by workers must involve training, maintenance of good health, and proper hygiene. Fortunately, implementation of GAPs worker health and hygiene training programs can be done without major equipment purchases or capital investment. It does require company commitment and time but offers significant risk reduction for the amount of money invested.

A significant challenge that currently exists for the implementation of worker training programs is the apparent lack of understanding of the importance of worker hygiene by growers and packers. Through a statewide survey of over 450 growers and packers in New York, it was learned that many growers did not have a worker training program related to health and hygiene. When asked, "Do you offer worker training that specifically addresses the importance of hand washing and personal hygiene?" 57.1% of those responding said no. When they were asked why they had not implemented such a program and were given a list of eight optional responses, the most frequent response (29.8%) was that "the workers are not interested." This response was not supported by results of a parallel survey conducted with more than 680 New York farm workers. When farm workers were asked, "Would you be interested in receiving information and training on proper hand washing for your own protection and to protect the fruits and vegetables you harvest and pack?" 73.7% of the respondents said yes. Growers also were given the option to provide their own response to why they did not offer training, and many did. Some of the responses that reveal the existing challenges included "not needed at my operation," "common sense information," "family workers," "they should have already learned the importance of this; not my job to raise workers," "not a high priority," and "I am not in the hygiene business." These responses indicate that many growers do not understand the link between food safety, worker health and hygiene, and produce handling or that they are not specifically concerned with this link. It is also possible that they do not associate the commodity they grow with a ready-to-eat food, but we know that many commodities are hand harvested and directly packed into consumer-ready containers. Many commodities are not washed prior to reaching the consumer because this would have a negative effect on postharvest quality, and it is known that many consumers eat fresh produce without washing the commodities at home. There is evidence to suggest that postharvest washing is not always effective at removing pathogens (see chapter 4 for an in-depth discussion).

Gastrointestinal viruses such as norovirus, which are commonly identified in foodborne outbreaks, can be spread over a wide area in aerosol droplets from a vomiting episode (40). Consider the possibility that an ill worker infected with norovirus experiences a sudden onset of symptoms and vomits in a field that is being harvested, spreading viral particles throughout the field. This hypothetical situation demonstrates the overall importance of workers reporting illnesses and of awareness of how illness can impact the safety of fruits and vegetables.

GAPs that should always be addressed in worker training programs include proper hand-washing procedures, practices involving worker illness, and proper use of sanitary facilities, including toilets. Many growers have noticed that workers do not always put used toilet paper in the toilet. In the survey of New York farm workers that was mentioned above, when asked, "When you use the toilet, where do you put the toilet paper?" of the respondents, 46.6% said in the toilet, 44.8% said in a trash can, 1.6% said next to the toilet, 0.9% said on the ground, and 6.2% provided their own responses, which included "in a bag to throw away later," "in the woods," "I go in the field and leave the paper there," and "I can't answer—there are no toilets." These responses do not include every independent response but represent a few of the diverse comments that were received. The belief that toilet paper belongs in the trash likely originates from the fact that many farm workers immigrate from countries where the plumbing systems cannot tolerate the deposition of toilet paper directly into the toilet. It is common in foreign countries to dispose of used toilet paper in a can near the toilet and not directly into the toilet. The responses to this question clearly indicate the need to clarify precisely how toilets in the United States work and farm expectations for sanitary practices, as well as the need to provide sanitary facilities for workers.

GAPs principles can be incorporated into worker training programs that already exist, such as those that discuss pesticides and proper harvesting techniques. The important thing is that workers are specifically told that they are expected to practice proper hygiene, including hand washing, because they are in contact with a ready-to-eat food product. Proper hygiene not only protects the products that are harvested and packed but also directly impacts health. Most farm workers are paid by the hour or by piece rate and are usually hired seasonally; protecting their health allows them to protect their income. It is a matter of personal decency and privacy to provide clean, sanitary toilets and hand-washing facilities. Regardless of the legal requirements in individual states, if a grower has one farm worker out in a field for any length of time, there should be a clean toilet and hand-washing facilities that are properly stocked with soap, water, and single-use paper towels. Drinking water also should be provided to ensure that workers do not get dehydrated and ill while working. If management or crew leaders see workers urinating or defecating in the field instead of in the provided facilities, they need to let the workers know that this is not appropriate field behavior. Discussing urination and defecation is often difficult, but if the importance has been emphasized in worker training programs, management can simply refer to previous training. Providing clean, sanitary, well-stocked hygiene facilities and enforcing proper use will promote good health and hygiene, reinforcing the farm's commitment to food safety.

IRRIGATION WATER QUALITY

Irrigation water quality plays a significant role in produce food safety, but currently in the United States there are no irrigation water quality regulations. In the past, the focus on irrigation water quality meant mineral content, hardness, pH, and other factors that affect plant health. With the focus changing to include food safety, microbial quality is now part of the irrigation water quality discussion. Water used for irrigation can be of variable quality, ranging from potable to surface water from sources such as rivers to treated and untreated municipal water. It is also delivered to the plants by both overhead and surface (drip, trickle, and subirrigation, etc.) methods. Many of the water quality recommendations are based on reclaimed or reused water, and the standards are quite variable. For instance, the U.S. Environmental Protection Agency (EPA) and U.S. Agency for International Development (USAID) in 1992 released guidelines for agricultural reuse of wastewater that suggested that there should be no detectable fecal coliforms per 100 ml of water in order for the water to be used for surface or spray irrigation of any food crop, including crops eaten raw (5). Brackett (6) suggested that "only clean potable water should be used for irrigation," but these standards may not be reasonable when viewed from a global perspective or from a production perspective. Potable water is an important and limited resource.

The *Canadian Water Quality Guidelines for the Protection of Agricultural Water Uses* recommends a maximum of 1,000 coliforms per 100 ml of irrigation water and 100 fecal coliforms per 100 ml of irrigation water (1, 46). The Ministry of Water, Land and Air Protection (Water, Air and Climate Change Branch) of the government of British Columbia established both quality standards and testing frequency standards. For water used to irrigate crops eaten raw, fecal coliform levels should not exceed 200 CFU/100 ml as a geometric mean, *E. coli* levels should not exceed 77 CFU/100 ml as a geometric mean, and enterococcus levels should not exceed 20 CFU/100 ml as a geometric mean (52). For all of these quality standards, the ministry stipulates that there should be at least five samples tested in a 30-day period (52). The World Health Organization (WHO) guidelines for the use of treated water in agriculture for crops likely to be eaten uncooked recommend a geometric mean of ≤1,000 fecal coliforms per 100 ml of water and an arithmetic mean of ≤1 intestinal nematode egg per liter of water. To achieve these levels, the WHO recommends a series of stabilization ponds or equivalent treatment that will result in the recommended quality (5). There are no standards for virus content, and only the WHO includes standards for intestinal nematodes, although it has been documented that irrigation water in the United States and Central America often contains microsporidia such as *Giardia* and *Cryptosporidium* spp. (49).

In considering these variable recommendations, it would be helpful to know the risk of human illness that may result from following a particular set of standards. Shuval et al. (41) developed a risk assessment model attempting to determine what the impact of these variable recommendations would be on human health. They submerged cucumbers and long-leaf lettuce in water and determined the amount of water that persisted on each commodity. Using this water volume, they took both the WHO and the EPA-USAID water recommendations and applied a modification of the Hass et al. risk assessment model for drinking water (20a). Both sets of recommendations were associated with lower risk of illness than is presently deemed acceptable by the EPA in relation to microbial standards for drinking water (1 case/10,000 persons/year). Shuval et al. (41) also estimated the unit cost of treatment of wastewater to meet the various guidelines and found that the adoption of the EPA-USAID standards would result in the expenditure of $3 to 30 million per case of disease prevented (41). So the higher water quality standards will prevent more illnesses but at a significant cost, and the prevention measures are significantly beyond present EPA standards for drinking water.

The overall irrigation demand in the United States is also a consideration when the possibility of implementing water quality recommendations is evaluated. Is it possible to use potable water to produce all horticultural crops that are to be consumed raw? The 2002 farm census data indicate that approximately 52 million acres of farmland are irrigated annually in the United States. Roughly 51% of the irrigation water is delivered through overhead sprinkler systems, and the remaining 49% is delivered by trickle, subirrigation, or gravity flow (http://www.nass.usda.gov/census/). The use of surface water to meet irrigation needs throughout the country is a prevalent practice. In states where water rights are limited and water availability is scarce, is it wise to suggest the use of potable water for irrigation when there may not be enough to meet drinking water needs?

Given that this chapter is focused on GAPs for fruit and vegetable production, it is important to consider the impact that current EPA-USAID water quality recommendations could have on U.S. producers. If fresh fruit and vegetable producers are prohibited from irrigating with surface water or water that is of poorer quality than presently recommended, how many growers will stop growing fresh produce? After a survey of New York fruit and vegetable producers, Rangarajan et al. (32) reported that 37% of the survey respondents used streams to irrigate and 50% irrigated from ponds. The respondents to this survey represented 36% of the produce acreage in New York. Although this is just one state, the results highlight the impact that would occur if the present recommendation had to be followed. If fruit and vegetable growers stopped producing fruits and vegetables, the supply would decrease but not the demand. This demand could be met by imported produce, but to be fair to U.S. producers, the imported commodities would have to be produced with water meeting the EPA-USAID standards. This scenario moves the issue beyond produce food safety and into the competitiveness of U.S. producers. These comments are not meant to advocate one water quality standard over another, but in considering irrigation water quality, these are a few of the issues that need to be fully discussed.

Irrigation Methods

Irrigation methods can greatly influence the risk of contamination of fresh fruits and vegetables while they are in the field. Irrigation methods that deliver water directly to the soil without wetting the plant or the ripe fruit or vegetable are lower risk than those that wet the entire plant (44). Drip, trickle, and subirrigation are examples of surface irrigation methods. Although there has been research that indicates that pathogenic organisms can travel internally through the plant, this study used larger inoculums than would naturally occur and was done in an environment where there was no natural soil competition (45). It is important to consider that pathogens may be able to move internally into a plant because many microorganisms are pathogenic to humans in extremely low numbers and internalization would offer protection from any type of postharvest decontamination step. Currently it is impossible to determine the actual risk to human health since the studies have not used parameters reflective of field conditions. Many growers who have switched from overhead to surface irrigation have noticed increases in crop yield as well as reduction of certain plant diseases. Although surface irrigation lowers the risk of contamination from irrigation water, many farms are not prepared financially or practically to install surface irrigation equipment due to the magnitude of acres being farmed or crop-specific issues such as growing practices or prices received.

Sometimes water is used not solely for irrigation but also for crop protection. Water is used for protection from frost for many cold-sensitive crops. Usually this water is applied prior to fruit ripening, so the risk is low. Water is also used to mix topical protective sprays. The risks associated with this type of water use are dependent on what is mixed with the water. Many protective sprays (fungicides and insecticides, etc.) have antibacterial qualities and when mixed with nonpotable water inhibit bacterial growth or kill existing bacteria that may be in the water. However, some may promote growth especially if left in the mixing tank for extended periods of time (20, 51). Levels of other pathogenic microorganisms such as parasites are not significantly reduced, and sporulation and viability are usually not significantly impacted, depending on the pesticide (38). Since the data regarding pathogen survival are different for each topical spray, making generalizations is difficult. It is also difficult to determine the fate of pathogens that may be in topical sprays once they are applied and subjected to the sun, high temperatures, and various moisture conditions. The real concern is whether or not pathogens survive on fruits and vegetables destined to be consumed.

In preliminary studies, selected horticultural commodities grown using overhead irrigation and topical sprays mixed with nonpotable water showed no end-product contamination with *Salmonella* spp., *E. coli* O157:H7, or *Listeria monocytogenes.* Most contamination is sporadic and difficult to find using standard sampling protocols since the amount of product sampled is so small compared to the amount grown. An interesting project was conducted in Norway in which 475 samples of various fruits and vegetables were analyzed for parasite contamination and 6% were found to be contaminated with either *Cryptosporidium* oocysts or *Giardia* cysts (35). All samples were collected postharvest either through commercial produce distributors or through local retailers. These items were not linked to outbreaks and included imported as well as domestic produce. Discovery of contamination usually occurs following a human foodborne illness outbreak, but in this case random sampling revealed low levels of parasitic contamination.

Present GAPs Recommendations for Irrigation

- If surface water is used, it should be tested for *E. coli* on a regular schedule to monitor microbiological quality and any changes that may occur due to unusual contamination events. If water test results indicate a contamination event, attempts should be made to identify the cause and water should not be applied to ripe crops. Water testing records should be kept on file.
- Drip or surface irrigation should be used when possible to prevent direct wetting of the plant or ripe fruit or vegetable. This also reduces plant disease.
- Potable water should be used for mixing topical sprays.
- Producers should be active in local watershed management and be aware of factors influencing their watersheds.
- If well water is used, producers should be sure that the well is capped and properly constructed. Wells should be tested at least once a year to monitor microbiological quality.

MANURE USE AND LAND SELECTION

Manure used as a soil amendment provides many benefits to the soil. It increases the organic carbon and overall organic matter that, in turn, improves granulation, water infiltration, water-holding capacity, nutrient content, soil biota activity, soil fertility, soil tilth, and overall productivity (12). It also serves the animal production industry as a way to manage animal waste. Unfortunately, many animals shed pathogenic organisms in their manure and these organisms can be transferred to fruits and vegetables grown in the field.

Abandoning manure use as a way to reduce risk is one option but not necessarily the best when all the benefits of manure use are assessed. Understanding the risks that exist and minimizing these risks through GAP implementation are one reasonable approach.

Animal manure from cattle, sheep, swine, and chickens can contain pathogenic microorganisms such as *Salmonella, E. coli* O157:H7, and *Campylobacter jejuni* (3, 48). Several herd management strategies are being researched to determine their effectiveness at reducing the prevalence or influencing the shed of pathogenic organisms in manure. These strategies include the use of vaccinations and feed additives, diet shifts from grain to grass or hay, and the use of probiotics and bacteriophages (39).

It seems unlikely at this time that pathogenic microorganisms will ever be completely removed from raw manure, so risk-minimizing strategies often must be implemented to affect the timing of manure use, the application of the manure, or the creation of compost. The National Organic Standards recommend a 120-day interval between the application of raw manure and harvest for crops whose edible portion has direct contact with the soil and a 90-day interval for crops whose edible portion does not come into contact with the soil (http://www.ams.usda.gov/nop/NOP/standards/ProdHandPre.html). Present standards from the National GAPs Program recommend applying raw manure at least 120 days prior to harvest of the crop, making no distinction for the type of crop. Other research suggests manure application at least 210 days prior to harvest, noting that the data are specific to carrots and onions grown in Georgia (25). The Georgia study highlights a few key points. Pathogen survival is often dependent on the soil type, competition with the natural soil flora, soil temperature, and the crop that is being grown (37). It also depends on the pathogen that is being evaluated. Pathogens such as protozoan parasites are often difficult to detect in low numbers and have different behavioral and survival profiles in the environment than bacteria and viruses (21).

Many growers are not pleased with the present recommendation to apply raw manure 120 days prior to harvest, particularly growers in northern locations who have very short growing seasons and have animals that produce manure on a daily basis. Managing the 120-day interval requires significant amounts of land as well as land rotation. Manure applications influence not only produce safety but also environmental safety, since poorly planned applications can result in runoff and contamination of rivers, lakes, and streams, impacting local watersheds. As with any issue, the more research data available to support the recommendation, the better implementation and adoption will be.

Proper application of manure and waiting periods are important for produce safety, but sometimes fear of the unknown results in industry requirements that are not based on science. Some produce buyers have implemented buying requirements that demand that manure not be applied for 5 years prior to harvest or, in some extreme cases, that land never have had a manure application. The last requirement is almost impossible to meet since most of the farm land has been farmed for decades. We found no data on manure and pathogen survival that could address the 5-year application standard, but it is clear that this type of requirement will impact growers and the fresh produce commodities they produce. The option of composting raw manure may be one way to meet industry demands and still utilize raw manure.

Proper composting is an effective way to reduce the pathogen load in manure prior to application. It needs to be stressed that *proper* composting is effective. Many growers believe that aging manure or simply allowing it to sit in a pile for extended periods of time constitutes composting. Though time will affect the pathogen load in manure, this method is not a consistent way to kill pathogens. The National Organic Standards final rule requires that compost be produced through a process that results in an initial C:N ratio of between 25:1 and 40:1 when plant and animal materials are combined. If the compost is produced using a windrow system, the materials must maintain a temperature of between 131°F and 170°F for 15 days, during which time the materials must be turned a minimum of five times (http://www.ams.usda.gov/nop/NOP/standards/FullText.pdf). There are other composting descriptions that can be found in the National Organic Standards, but additional research has supported the above-mentioned methods as being effective at reducing pathogens (29).

Management challenges to composting raw manure include land availability, maintenance, and a method to prevent recontamination by both domestic and wild animals. If compost piles do not meet the minimum heat requirement but linger in the mesophilic temperature range, the situation can actually result in pathogen growth (50), so proper monitoring is critical. If purchasing compost from other farms, growers are encouraged to get documentation that states how the manure was composted as well as reporting any tests that were done regarding its microbial quality.

SANITATION

Field Sanitation

Sanitation is a part of GAPs where significant control can be applied both in field operations and in the packinghouse. Much information is available about sanitation, but this section will focus on key elements that specifically pertain to implementation of GAPs on the farm. The obvious sources of pathogens have already been discussed and include contaminated irrigation water, unsanitary equipment and tools, wild and domestic animals (and their manure), insects, and workers with poor health and hygiene.

All harvest equipment, including harvest totes, baskets, bags, and bins, should be constructed of easily cleanable materials. This equipment should be cleaned and sanitized between daily uses or as appropriate given the harvest schedule. Many older bins are constructed of wood, which is difficult to clean. When purchasing new equipment, growers should keep sanitation in mind and purchase equipment that is easy to clean and sanitize. For products that are field packed into single-use containers, the containers should be clean and stored in sanitary conditions to prevent contamination prior to use. Worker health and hygiene are critical at this point since microorganisms can be easily spread from contaminated hands to harvested crops as described earlier in the chapter.

Packinghouse Sanitation

The packinghouse should be properly designed and constructed, protected from the environment, and maintained in a clean and sanitary manner. Several produce-associated outbreaks have been traced back to poor sanitary conditions in packinghouses, and these conditions were thought to contribute to the outbreaks (22). Some packinghouses are open to the elements, which makes excluding insects, rodents, birds, and dust very difficult, but there are sanitary practices that can be implemented in any packinghouse situation.

Pest Management

Three important objectives for addressing pest management in the packinghouse are as follows:

1. Preventing the entry of insects, rodents, and birds.
2. Restricting the availability of shelter and food.
3. Destroying pests if they do gain access to the facility.

The key to effective pest management is prevention through good sanitation and attention to detail (18). A pest management program begins with a well-thought-out plan to ensure that all possible sources of entry of insects, rodents, and birds are addressed. There should be no openings around outside doors, especially at the bottom, and these doors should all have tightly fitting sweeps and seals. All cracks and crevices in walls and doors and openings of 1/4 in. (0.6 cm) or less in size, which may provide entrance to pests, should be properly sealed. Windows and screens should be tight fitting and in good repair, and air curtains, if used, should be functioning properly. All openings around pipes should be permanently sealed with an adequate protective material such as concrete, caulking, or other effective products. Since rodents can also gain entry by climbing pipes or wires, metal shields or guards should be placed over these. In addition, vents and exhaust fans, etc., should have tight-fitting, self-closing louvers or screens to exclude insects and rodents.

The outside wall of the building, especially at the junction of the wall and the ground, should be properly rodent proofed to deter rodents from burrowing under the wall. A rodent control program with tamper-resistant poison bait stations can be set up around the building perimeter or fence line to deter rodents from entering the facility. Although new technologies are available, the use of poison baits inside the facility is not recommended. Spring traps, glue boards, and mechanical traps can be used inside the facility to trap any rodents that do gain entry. For birds, the use of ledges and overhangs on buildings should be avoided. Bird netting and other physical devices can be used in already existing areas of the building to discourage birds from roosting, loafing, or nesting (18). To prevent contamination by bird feces, discouraging the presence of birds is of particular importance in coolers and packinghouses where harvested product is stored.

All outside doors (especially receiving doors) should be kept closed when not in use. Grass, ornamental shrubs, and other landscaping should be kept away from the building perimeter and should be properly maintained to avoid harborage and hiding places for insects, rodents, and reptiles. The exterior of the packinghouse should be kept neat and clean. Trash, including old equipment, cartons, pallets, boards, and any unwanted materials that might be a breeding ground for insects and rodents, should be removed (18). Spilled, damaged, or rotting produce should be picked up quickly and disposed of properly. Garbage management in and around the packinghouse is an important aspect of pest control and management. Garbage should be kept in secured, closed dumpsters away from the building and emptied or removed from the facility frequently, especially during the summer months or periods when temperatures are very warm.

Design and Construction

The packinghouse should be constructed using the principles of sanitary design. Inside the building, smooth materials impervious to dirt and moisture should be used as much as possible to minimize product contamination and facilitate cleaning and sanitizing procedures. Attention should be given to plant layout as well as to the flow of product and personnel through the facility. The construction of the floors, walls, and ceilings as well as the placement of equipment, hand-washing sinks, employee facilities (locker rooms, toilets and lunch rooms), floor drains, and other items should be carefully considered during building construction or renovation. Equipment should be installed with thought given to its cleanability and its distance from walls and other equipment, as well as to how it will be mounted to the floor, wall, or ceiling.

Lighting in the packing areas of the building should consist of a minimum of 50-foot candles (540 lx) so that employees have adequate light to work safely (18). All lights should be properly shielded to protect the product in the event of breakage.

Equipment and Food Contact Surfaces

Many fruits and vegetables are considered ready-to-eat products, so every surface that they contact should be clean and sanitary. Packinghouse workers should think of all packing surfaces as food contact surfaces. The surfaces of conveyors, sorting and grading tables, and other equipment that comes in contact with produce can accumulate plant residues, debris, plant juices, soil, and other materials that can become a source of contamination. Biofilms, which are an accumulation of bacterial cells and food debris attached to food contact surfaces, can easily form and develop if cleaning and sanitizing practices are not performed frequently or thoroughly. Once formed on surfaces, biofilms are difficult to remove. Figure 2 shows an electron micrograph of a stainless steel surface with microorganisms attached. This image illustrates the point that proper cleaning and sanitizing procedures need to be performed to remove these organisms from food contact surfaces. All equipment on the packing line, as well as floors, floor drains, lunch rooms, and restrooms, should be thoroughly cleaned and sanitized daily. Depending on use, other areas of the facility may be cleaned and sanitized less frequently.

It has been shown that microorganisms can survive on equipment and environmental surfaces including brushes, rollers, wash and spray equipment, conveyor systems, floors, and floor drains as well as in other areas in the facility, so a uniform schedule of cleaning and sanitizing (a master schedule) must be developed, implemented, maintained, and properly recorded. A sanitation record-keeping system provides evidence that cleaning and sanitizing procedures were properly carried out. There are a wide variety of cleaners and sanitizers on the market today, so it is recommended that growers and packinghouse operators contact chemical suppliers for specific advice on their cleaning and sanitizing needs.

Employee Facilities

Since street clothes and personal belongings can contaminate produce, equipment, and surfaces, it is important that workers have a clean, secure location, such as a locker room, to store personal items. Workers also should

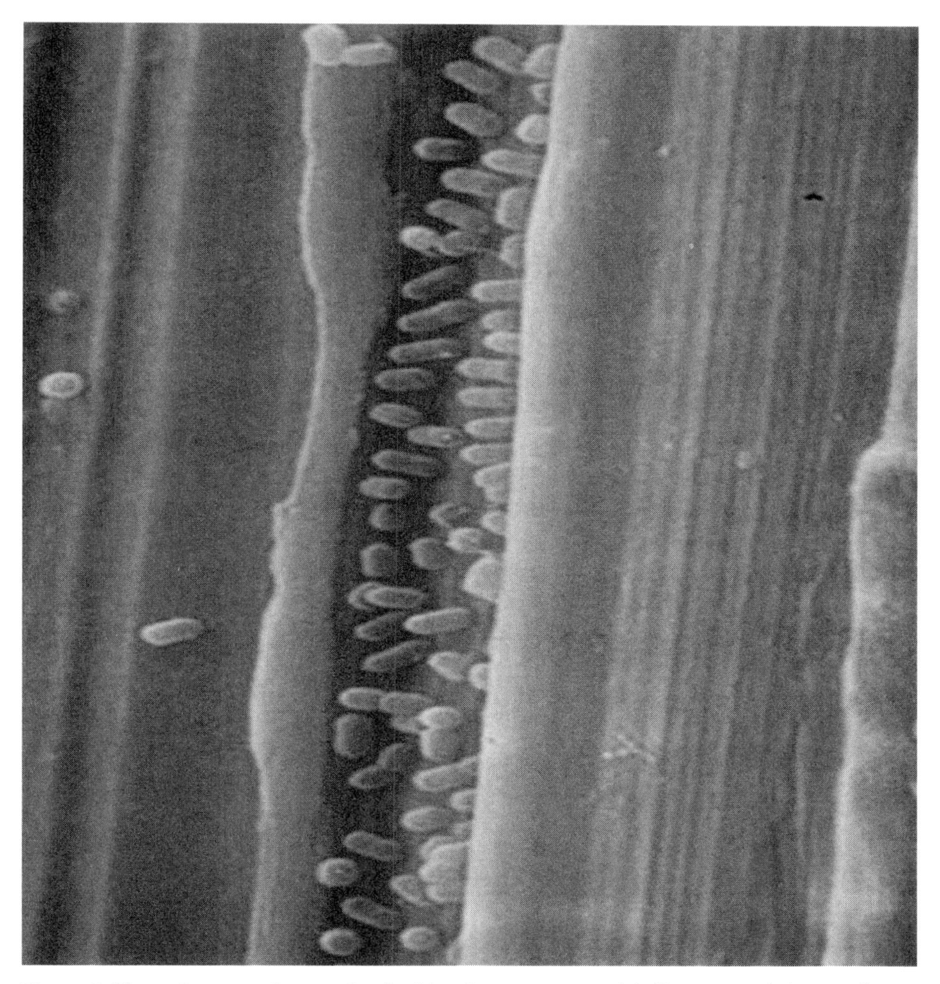

Figure 2 Photoelectron micrograph of a *Listeria monocytogenes* biofilm on a stainless steel surface. Proper cleaning and sanitizing are required to remove microorganisms from a food contact surface. (Image used with permission from Amy Wong and ASM Biofilms Collection.)

have a clean, comfortable, and pleasant place where they can store food they bring from home, take breaks, and eat lunch. Locker rooms should be kept clean and orderly and be located in an area where contamination of produce and equipment cannot occur. Toilet facilities in the packinghouse should be kept clean, in good repair, and properly supplied with toilet paper, hot water, soap, and either single-use paper towels or hot-air dryers. The cleanliness of the lunch room, locker room, and toilet facilities can affect the attitudes, work habits, and motivation of employees (18).

Hand washing is one of the most effective ways to limit the spread of bacteria or viruses from people to fruits and vegetables. Hand-washing sinks strategically placed throughout the packinghouse allow workers to properly wash their hands during their daily activities. Although it is a simple task, few people wash their hands frequently enough or properly. The proper hand-washing technique for people working with food is as follows.

1. Allow enough time to wash hands properly. Use water as hot as the hands will comfortably tolerate—at least 105°F.
2. After moistening the hands and exposed portions of the arm, soap thoroughly.
3. Rub the lathered hands together for at least 20 s (sing "Happy Birthday" twice or the alphabet song once).
4. Pay particular attention to the areas around and underneath the fingernails and between the fingers and exposed areas of the forearm.
5. Rinse thoroughly with clean water.
6. Dry hands using a single-service, disposable towel or a hot-air dryer.

Hand washing is an activity that should always be done before beginning work and should be repeated frequently throughout the day. It is especially critical after performing any of the following activities:

- Using the toilet
- Coughing, sneezing, or using a tissue or handkerchief
- Handling raw foods, particularly meat and poultry
- Touching areas of the body such as the mouth, nose, hair, and ears or scratching anywhere on the body
- Touching infected or otherwise unsanitary areas of the body
- Touching or handling soiled materials, equipment, or work surfaces
- Handling garbage
- Eating foods or drinking beverages
- Smoking or using chewing tobacco
- Returning to the work station from breaks or lunch

In many packinghouses, workers use disposable or reusable rubber gloves when handling produce. Clean, intact gloves can provide an effective barrier between hands and produce. Sometimes workers forget that gloves also can become soiled and dirty, just like their hands, and can contaminate the produce that they are sorting, grading, and packing. Disposable gloves should be changed frequently, especially when they are worn for long periods of time; get ripped, damaged, or soiled; or are used for tasks other than product contact. Once disposable gloves are taken off, they should be discarded and not used again. Reusable gloves should be washed and sanitized frequently and thrown away when they become old, soiled, torn, or uncleanable. Gloves are not a substitute for proper hand washing. The proper procedure for glove use is to wash hands properly and then put on clean, intact gloves. For some workers, the use of gloves inhibits their job performance due to lack of tactile sensitivity. In these instances, frequent and proper hand washing followed by use of a sanitizer hand dip (like an iodophor) is the best way to reduce microbial risks.

Water Quality

The microbiological quality of the water used is very important to the safety of the fruits and vegetables. Irrigation water quality was covered earlier in this chapter, so this section on water sanitation will focus on postharvest water uses. Potable water should be used for all postharvest applications, including cooling and cleaning. Many operations recirculate water through the cooling and cleaning systems, which include dump tanks, flumes, hydrocoolers, and spray washers, to conserve water and energy (47). As a result, dirt and organic matter including plant material and disease-causing pathogens can accumulate in the water. In order to prevent the transfer of disease-causing organisms from the water to produce, as well as from one produce item to another, postharvest water should be disinfected. Disinfection will inactivate or destroy many human and plant pathogens, resulting in a safer product with improved shelf life. Disinfection can be achieved using chemicals such as chlorine, iodine, ozone, and peroxide or physical processes such as microfiltration and ultraviolet illumination (47). Some pathogens are not destroyed with chemicals such as chlorine, so the pathogens of interest should be a consideration when a disinfection strategy is chosen. If a chemical sanitizer is used, the levels need to be monitored and maintained in order to ensure proper disinfection. These tasks can be done manually or automatically. Automated sanitizer-metering systems that maintain the level of sanitizer strength and adjust pH to allow for maximum sanitizer efficacy should be properly calibrated, checked regularly throughout the day, and then checked with manual methods to ensure that the sanitizer strength is correct and that the system is functioning properly. Physical processes (microfiltration and ultraviolet illumination) also require monitoring to ensure proper function and effective disinfection.

A well-designed, properly constructed, and well-maintained and -managed packinghouse can significantly reduce microbial risks for fresh fruits and vegetables.

RECORD KEEPING

One of the most significant challenges to GAPs implementation is the development of records that document adoption and maintenance of a farm food safety plan. Many practices that are now termed GAPs have been used in the agricultural industry for many years. There is overlap with "best management practices" (BMPs) and "agricultural environmental management" (AEM) practices, although the focus of GAPs has been specific to reducing microbial risks associated with food safety. Although many farms have implemented GAPs, they often have not documented what they have done and often do not keep up-to-date records on issues like field toilet sanitation, worker training programs, manure application rates and dates, and other pertinent information. After conducting a survey of fruit and vegetable growers, Rangarajan et al. (34) identified three key areas in which small farms were specifically in need of training: record keeping for manure applications, composting processes to achieve pathogen kill, and sanitation of wash water. All records need to be kept on the farm for reference, for use in GAPs audits, and for emergencies.

Record keeping is a key component of any GAPs farm food safety plan. Implementing record keeping can be time-consuming, but the critical time investment is in the design of the system and the training to make certain that the record keeping occurs at the right time. Recording progression of GAPs implementation does not require the development of new record-keeping systems and often can be achieved through modification of current farm records. Many record-keeping issues, such as recording cooler temperatures, can be achieved by placing a clipboard in the cooler and recording the temperatures at the beginning, middle, and end of each day. Other GAPs records, such as those on rodent control and regular sanitation of field toilets, can be provided by the companies contracted to perform these tasks. Growers who are the most burdened by record keeping are those who have extremely small operations, since most of the farm management falls to one person and it is likely that this one person is already busy with other farm business. Record keeping is critically important if a farm decides to have a third-party audit. Auditors depend on records to confirm implementation and day-to-day maintenance of a GAPs program. If not participating in a third-party audit, some growers may feel that record keeping is not pertinent to their operations, but should a problem arise, record keeping provides evidence that a task was performed.

Record keeping is essentially the only way to show due diligence, which is defined as the attention and care legally expected of a person or business. If an outbreak were to occur, record keeping, trace-back strategies, and crisis management would be the most utilized parts of a farm operation. Crisis management will be discussed later in the chapter. Nothing can offer ultimate legal protection, but due diligence is definitely better than negligence, and the expectation is that growers will be following standard growing practices such as GAPs. Since there have been many documented foodborne illnesses associated with fresh produce, it would be difficult for growers to deny that food safety is an issue. Knowing that it is an issue means that growers are expected to have implemented practices such as GAPs to reduce risks where possible, and the only way to show implementation is through an effective record-keeping system.

One of the most important types of record keeping for fresh fruit and vegetable growers is product identification and a system for product recall and trace back. Each piece, container, or lot shipped from the farm should be coded to allow farm management to trace it from the field of origin to the distributor. These coded lot numbers should be included on the bill of lading. In addition to coding lots of product, each farm should have a written recall plan that includes the name of an employee who serves as a recall team leader; a process for notification of important contacts, including the customer, the public, and regulatory agencies; procedures for implementing the recall; strategies for handling and disposing of recalled produce; and methods for verifying the recall plan's effectiveness (33). Mock recalls should be conducted to see how the recall plan works and identify any weak areas so that they can be corrected prior to a trace-back event. It is also important that the recall plan be updated regularly to keep phone numbers for buyers, distributors, farm support agencies, and other critical contacts updated, as well as to record any changes that occur in the operation over time (33).

BEYOND THE FIELD: GAPs FOR FARM SECURITY

Two areas that are closely linked but are often not thought of as being specifically part of GAPs are biosecurity (food protection and defense) and crisis management. A biosecurity plan covers the intentional contamination of the operation, staff, or produce with a chemical or biological agent. The risk for individual farms is impossible to predict, but there are GAPs that can be implemented to make any operation more secure. The most obvious of these is securing farm buildings by locking doors and windows when buildings are not being used, particularly if they house coolers containing harvested commodities. Locking chemical storage buildings and limiting access to keys that open any farm door is another safety step that can be taken. Controlling the flow of people on and off the farm, whether they are employees or visitors, and basic visual vigilance will enhance farm safety. The idea of biosecurity can be taken to additional levels with fences, cameras, and guards, etc., but for the vast majority of family farmers, this would be an expense that far outweighs the value. Larger operations that have significant volume from multiple farms have moved in the direction of more sophisticated security out of necessity and increased risk due to the sizes of the operations. Although every operation may not need a technologically advanced biosecurity plan, every operation should consider how changes can be implemented to strengthen farm security. Regardless of the measures taken to increase farm security, every operation should have a crisis management plan.

A crisis management plan is relevant for produce food safety as well as operational safety. Product recalls, foodborne illness outbreaks, accidental injury or death of a critical farm staff member, natural disasters, and power outages are just a few examples of situations in which a crisis management plan may be crucial to an operation. Identification of important phone numbers, the chain of command for staff, and critical farm operations that must be maintained as well as farm operations that can be ignored for short amounts of time should be addressed in the plan. Regardless of size, every operation needs to have a plan to assist the staff in successfully navigating a crisis with as little negative effect as possible.

Specialty Produce Markets

Many growers may utilize specialty markets to sell their commodities outside of the wholesale market. These growers still need to be aware of GAPs and will need to evaluate their own operations to determine if market-specific GAPs need to be implemented. For instance, some growers may participate in farm markets where consumers are given samples to taste. Cutting and serving fresh produce present other opportunities for contamination. Rinsing the product with cool water, preparing it using knives and cutting boards that have been cleaned and sanitized, storing samples in a clean container that is kept on ice or refrigerated, and disposing of any samples that have not been consumed after 4 h are just a few examples of GAPs that would be specific to this situation.

Other specialty markets that present their own challenges are "pick-your-own" ("U-pick") operations, direct farm markets, and entertainment farms that include petting zoos. Most of these require direct interaction with customers, and GAPs implementation in these environments can be difficult because customers do not always understand agriculture or the risks associated with fruit and vegetable production. One good example is that people do not always wash their hands after using the bathroom, not to mention after petting friendly farm animals at a petting zoo. Recently, there have been several outbreaks attributed to petting zoos. Even though a farm may provide hand-washing sinks and post signs requesting that patrons wash their hands after visiting the exhibit, enforcing this behavior is not easy; still, it is critical to the safety of the patrons and the operation. The best approach for any niche- or specialty-market grower is to evaluate the operation and implement practices that reduce the identified risks. Again, the uniqueness of each operation makes it difficult to address all the possible situations that may occur, but the growing popularity of these markets makes them worth mentioning.

HISTORY OF FOOD SAFETY AND THE ROLE OF GAPs

"GAPs" is the appropriate acronym to be using, and these practices are the appropriate actions to be discussing when talking about food safety on the farm. Unfortunately, it is not uncommon for produce buyers to ask fruit and vegetable growers to provide an HACCP plan for their field and packinghouse operation. This request often results in panic and confusion when growers read what an HACCP plan entails and attempt to determine how they would control items that they feel would be categorized as critical control points. Even among the scientific community there is debate and discussion, with many feeling strongly about what words and acronyms are and are not appropriate. This is a matter of semantics, but an important one since, in a scientific discussion, it is extremely important to be clear on the exact meaning and idea that are being conveyed. It is our opinion that GAPs are appropriate at the farm level due to the level of control that is achievable in the field.

In determining the role of GAPs in produce food safety, it is important to first discuss their context within the larger food safety arena. In the 1950s, as part of the developing human space program, there was an attempt to design a food safety program that could ensure food safety without destructive end-product testing. The result of this effort was the development of the HACCP concept. This concept has since undergone several refinements, but it is now the national and international standard used in processing plants and involves seven specific principles. These principles ensure that there has been an analysis to determine hazards and critical control points to control hazards that are likely to cause illness or injury to consumers. HACCP programs work because there is a level of control that can be maintained over materials, machines, and processes in food-processing plants, retail food stores, and food service operations. One of the foundations of an effective HACCP program is the implementation of GMPs. These practices are described in Title 21, Part 110, of the Code of Federal Regulation (11b).

Although the HACCP system has long been recognized as the most effective and flexible system for ensuring the microbiological safety of foods, there have been few attempts to apply the HACCP principles to the production of fresh fruits and vegetables. Several HACCP plans have been developed for sprouted seeds, shredded lettuce, and tomatoes (36), but complete validation of these plans has not yet been accomplished (30). The available data are insufficient to develop validated HACCP plans for most fresh produce items (31). In most cases, it would be extremely difficult if not impossible to implement a HACCP program in the field. There are far too many variables, including weather, wild animals, irrigation water, and soil, as well as several other factors that are not easily controlled.

GAPs are not federally regulated and can be modified to meet the needs of the diverse farms that produce fruits and vegetables throughout the United States. When GAPs are implemented, they reduce microbial risks during growing, harvesting, and packing of fruits and vegetables. Prevention of risks is hard to quantify, but reviewing the causes of past foodborne illness outbreaks helps to identify areas that have been problematic in the past and that will likely be impacted by preventive measures.

MOVING FORWARD

As produce food safety and the concept of GAPs reach all fruit and vegetable growers, big and small, every farm will develop a GAPs farm food safety plan that addresses the microbial risks that exist in each of these unique operations. Figure 3 highlights some of the areas that are likely to be part of any GAPs plan and includes additional areas that may be important to growers who sell to specialty markets. In the area of produce food safety, the more information that is available, the easier it will be to ensure the safety of fruits and vegetables through implementation of GAPs that effectively and efficiently prevent contamination. The implementation of GAPs does not occur in a vacuum but is closely linked to the farm economy, commodity supply, consumer demand, buyer requirements, commodity price, international markets, media attention, and changing microorganisms. Determining how to best balance these issues while producing a crop that makes production profitable is a very complex process. Additional research that defines the amount of risk associated with certain practices, such as the use of poor-quality irrigation water and the use of soil amendments such as manure, will further define GAPs and reveal how best to reduce risks on the farm. Practical field research also will provide information about the natural competition in the soil or the effects of UV radiation and solarization on reducing microbial risks in the field.

Growers and packers who are now aware of the importance of food safety will certainly play an important role as they make decisions about their

A Good Agricultural Practices Farm Food Safety Plan Should Include the Following Sections

- Irrigation Practices
- Manure Use
- Worker Health, Hygiene and Training
- Toilet and Hand washing Facilities
- Field and Packinghouse Sanitation
- Pesticide Use
- Animal and Pest Management
- Postharvest Handling
- Crisis Management
- Recall and Traceback
- Farm Biosecurity
- Record Keeping

Specialty and Niche Markets May Need to Add the Following Sections

- Direct Marketing
- Farm Market Protocols
- Pick Your Own/U-Pick Operations
- Petting Zoos Including Animal Health

Figure 3 Key components of a GAPs farm food safety plan.

operations. The purchase of new equipment or supplies will be made with safety in mind, and thus they will choose materials that are easier to clean and sanitize. The same priorities should be applied when deciding to expand an operation or packing line. Designing new facilities with sanitation in mind will mean choosing building materials that can be sanitized and designing buildings with proper ventilation, floor drains, walls, and ceilings that promote easy cleaning and sanitation. Necessity is the mother of all invention, and many farms have made unique modifications to existing equipment or fabricated their own equipment to meet their commodity requirements. In some cases, this creativity is applied in the field, such as the development of coring machines and stainless steel harvesting equipment that can be cleaned and sanitized in the field at the end of each harvest day.

As this discussion of GAPs and produce safety comes to a close, it is important to reiterate several important facts. Present technologies cannot eliminate all potential food safety hazards associated with fresh produce that will be eaten raw (16). The focus is on prevention. Many GAPs can be implemented with little capital investment. The biggest challenges are viewing fruits and vegetables as ready-to-eat food products and understanding which risks are most likely to occur for a specific commodity. As education continues and research provides more insight into these risks, preventing contamination on the farm through the implementation of GAPs will become more effective.

REFERENCES

1. **Anonymous.** 1999. Canadian Council of Ministers of the Environment, Canadian water quality guidelines for the protection of agricultural water uses, p. 2. *In Canadian Environmental Quality Guidelines.* CCME Publications, Winnepeg, Manitoba, Canada.

1a. **Beru, N., and P. A. Salsbury.** 2002. FDA's produce safety activities. *Food Safety Magazine* **2002**(February–March):14–19.

2. **Beuchat, L. R.** 1998. *Surface Decontamination of Fruits and Vegetables Eaten Raw: a Review.* WHO/FSF/FOS/98.2. World Health Organization, Geneva, Switzerland.

3. **Beuchat, L. R.** 2002. Ecological factors influencing survival and growth of human pathogens on raw fruits and vegetables. *Microbes Infect.* **4**:413–423.

4. **Beuchat, L. R., and J.-H. Ryu.** 1997. Produce handling and processing practices. *Emerg. Infect. Dis.* **3**:459–465.

5. **Blumenthal, U. J., A. Peasey, G. Ruiz-Palacios, and D. D. Mara.** 2000. Guidelines for Wastewater Reuse in Agriculture and Aquaculture: Recommended Revisions Based on New Research Evidence. WELL Study. London School of Hygiene and Tropical Medicine, Loughborough University, Leicestershire, United Kingdom.

6. **Brackett, R. E.** 1999. Incidence, contributing factors, and control of bacterial pathogens in produce. *Postharvest Biol. Technol.* **15**:305–311.

7. **Center for Science in the Public Interest.** 2001. *Outbreak Alert: Closing the Gaps in Our Federal Food-Safety Net.* Center for Science in the Public Interest, Washington, D.C.

8. **Centers for Disease Control and Prevention.** 1997. Outbreaks of *Escherichia coli* O157:H7 infection and cryptosporidiosis associated with drinking unpasteurized apple cider—Connecticut and New York, October 1996. *Morb. Mortal. Wkly. Rep.* **46**:4–8.

9. **Centers for Disease Control and Prevention.** 2002. Multistate outbreaks of *Salmonella* serotype Poona infections associated with eating cantaloupe from Mexico—United States and Canada, 2000–2002. *Morb. Mortal. Wkly. Rep.* **51**:1044–1047.

10. **Centers for Disease Control and Prevention.** 2003. Foodborne transmission of hepatitis A—Massachusetts, 2001. *Morb. Mortal. Wkly. Rep.* **52**:565–567.

11. **Centers for Disease Control and Prevention.** 2003. Hepatitis A outbreak associated with green onions at a restaurant—Monaca, Pennsylvania, 2003. *Morb. Mortal. Wkly. Rep.* **52**:1–3.

11a. **Code of Federal Regulation.** 2001. 21 CFR 120.1 (title 21, part 120, section 1). U.S. Government Printing Office, Washington, D.C., via GPO Access at http://www.gpoaccess.gov/cfr/retrieve.html (p. 261–275).

11b. **Code of Federal Regulation.** 2005. 21 CFR 110.3. U.S. Government Printing Office, Washington, D.C., via GPO Access at http://www.gpoaccess.gov/cfr/retrieve.html (p. 220–230).

12. **Committee on the Role of Alternative Farming Methods in Modern Production Agriculture, Board on Agriculture, and National Research Council.** 1989. *Alternative Agriculture.* National Academy Press, Washington, D.C.

13. **Diez-Gonzalez, F., T. R. Callaway, M. G. Kizoulis, and J. B. Russell.** 1998. Grain feeding and the dissemination of acid-resistant *Escherichia coli* from cattle. *Science* **281**:1666–1668.
14. **Donaldson, M. S.** 2004. Nutrition and cancer: a review of the evidence for an anti-cancer diet. *Nutr. J.* **3**:19.
15. **Doyle, M. P.** 2000. Reducing foodborne disease: what are the priorities? *Nutrition* **16**:647–649.
16. **Food and Drug Administration, U.S. Department of Agriculture, and Centers for Disease Control and Prevention.** 26 October 1998, posting date. *Guidance for Industry: Guide to Minimize Microbial Food Safety Hazards for Fresh Fruits and Vegetables.* [Online.] Food and Drug Administration, Washington, D.C. http://www.foodsafety.gov/~dms/prodguid.html.
17. **Gagliardi, J. V., P. D. Millner, G. Lester, and D. Ingram.** 2003. On-farm and postharvest processing sources of bacterial contamination to melon rinds. *J. Food Prot.* **66**:82–87.
18. **Gravani, R. B., and D. C. Rishoi.** 1998. *Food Store Sanitation,* 6th ed. Lebhar-Friedman Books, New York, N.Y.
19. **Gravani, R. B., and D. L. Scott.** 2001. *Developing Effective Education and Training Programs for Food Processing Plant Management and Staff,* 2nd ed. Department of Food Science Extension, Ithaca, N.Y.
20. **Guan, T. Y., G. Blank, A. Ismond, and R. van Acker.** 2001. Fate of foodborne bacterial pathogens in pesticide products. *J. Sci. Food Agric.* **81**:503–512.

20a. **Haas, C. N., J. B. Rose, C. Gerba, and S. Regli.** 1993. Risk assessment of virus in drinking water. *Risk Anal.* **13**:545–552.

21. **Herwaldt, B. L.** 2000. *Cyclospora cayetanensis:* a review, focusing on the outbreaks of cyclosporiasis in the 1990s. *Clin. Infect. Dis.* **31**:1040–1057.
22. **Hilborn, E. D., J. H. Mermin, P. A. Mshar, J. L. Hadler, A. Voetsch, C. Wojtkunski, M. Swartz, R. Mshar, M.-A. Lambert-Fair, J. A. Farrar, K. Glynn, and L. Slutsker.** 1999. A multistate outbreak of *Escherichia coli* O157:H7 infections associated with consumption of mesclun lettuce. *Arch. Intern. Med.* **159**:1758–1764.
23. **Institute of Food Technologists.** 2001. *Analysis and Evaluation of Preventive Control Measures for the Control and Reduction/Elimination of Microbial Hazards on Fresh and Fresh-Cut Produce.* Institute of Food Technologists, Chicago, Ill.
24. **International Fresh-Cut Produce Association.** 2001. *Food Safety Guidelines for the Fresh-Cut Produce Industry,* 4th ed. International Fresh-Cut Produce Association, Alexandria, Va.
25. **Islam, M., M. P. Doyle, S. C. Phatak, P. Millner, and X. Jiang.** 2005. Survival of *Escherichia coli* O157:H7 in soil and on carrots and onions grown in fields treated with contaminated manure composts or irrigation water. *Food Microbiol.* **22**:63–70.
26. **Kaneko, K.-I., H. Hayashidani, K. Takahashi, Y. Shiraki, S. L. Wongpranee, and M. Ogawa.** 1999. Bacterial contamination in the environment of food factories processing ready-to-eat fresh vegetables. *J. Food Prot.* **62**:800–804.
27. **Litwak, D.** 1998. Is bigger better? *Supermarket Business* **53**:79–88.
28. **Morris, J. G., and M. Potter.** 1997. Emergence of new pathogens as a function of changes in host susceptibility. *Emerg. Infect. Dis.* **3**:433–441.
29. **Mukherjee, A., D. Speh, E. Dyck, and F. Diez-Gonzalez.** 2004. Preharvest evaluation of coliforms, *Escherichia coli*, *Salmonella*, and *Escherichia coli* O157:H7 in organic and conventional produce grown by Minnesota farmers. *J. Food Prot.* **67**:894–900.

30. **National Advisory Committee on Microbiological Criteria for Foods.** 1999. Microbiological safety evaluations and recommendations on fresh produce. *Food Control* **10:**117–143.
31. **National Research Council.** 2003. *Scientific Criteria To Ensure Safe Food.* National Academy Press, Washington, D.C.
32. **Rangarajan, A., E. A. Bihn, R. B. Gravani, D. L. Scott, and M. P. Pritts.** 1999. *Food Safety Begins on the Farm: a Grower's Guide.* National GAPs Program, Cornell University, Ithaca, N.Y.
33. **Rangarajan, A., E. A. Bihn, M. P. Pritts, and R. B. Gravani.** 2004. *Food Safety Begins on the Farm: a Grower Self Assessment of Food Safety Risks.* National GAPs Program, Cornell University, Ithaca, N.Y.
34. **Rangarajan, A., M. P. Pritts, S. Reiners, and L. H. Pedersen.** 2002. Focusing food safety training based on current grower practices and farm scale. *HortTechnology* **12:**126–131.
35. **Robertson, L. J., and B. Gjerde.** 2001. Occurrence of parasites on fruits and vegetables in Norway. *J. Food Prot.* **64:**1793–1798.
36. **Rushing, J. W., F. J. Angulo, and L. R. Beuchat.** 1996. Implementation of a HACCP program in a commercial fresh-market tomato packinghouse: a model for the industry. *Dairy Food Environ. Sanitation* **16:**549–553.
37. **Sadovski, A. Y., B. Fattal, D. Goldberg, E. Katzenelson, and H. I. Shuval.** 1978. High levels of microbial contamination of vegetables irrigated with wastewater by the drip method. *Appl. Environ. Microbiol.* **36:**824–830.
38. **Sathyanarayanan, L., and Y. Ortega.** 2004. Effects of pesticides on sporulation of *Cyclospora cayetanensis* and viability of *Cryptosporidium parvum*. *J. Food Prot.* **67:**1044–1049.
39. **Schamberger, G. P., R. L. Phillips, J. L. Jacobs, and F. Diez-Gonzalez.** 2004. Reduction of *Escherichia coli* O157:H7 populations in cattle by addition of colicin E7-producing *E. coli* to feed. *Appl. Environ. Microbiol.* **70:**6053–6060.
40. **Seymour, I. J., and H. Appleton.** 2001. Foodborne viruses and fresh produce. *J. Appl. Microbiol.* **91:**759–773.
41. **Shuval, H., Y. Lampert, and B. Fattal.** 1997. Development of a risk assessment approach for evaluating wastewater reuse standards for agriculture. *Water Sci. Technol.* **35:**15–20.
42. **Sivapalasingam, S., C. R. Friedman, L. Cohen, and R. V. Tauxe.** 2004. Fresh produce: a growing cause of outbreaks of foodborne illness in the United States, 1973 through 1997. *J. Food Prot.* **67:**2342–2353.
43. **Smith, J. L.** 1997. Long-term consequences of foodborne toxoplasmosis: effects on the unborn, the immunocompromised, the elderly, and the immuncompetent. *J. Food Prot.* **60:**1595–1611.
44. **Solomon, E. B., C. J. Potenski, and K. R. Matthews.** 2002. Effect of irrigation method on transmission to and persistence of *Escherichia coli* O157:H7 on lettuce. *J. Food Prot.* **65:**673–676.
45. **Solomon, E. B., S. Yaron, and K. R. Matthews.** 2002. Transmission of *Escherichia coli* O157:H7 from contaminated manure and irrigation water to lettuce plant tissue and its subsequent internalization. *Appl. Environ. Microbiol.* **68:**397–400.
46. **Steele, M., and J. Odumeru.** 2004. Irrigation water as source of foodborne pathogens on fruit and vegetables. *J. Food Prot.* **67:**2839–2849.

47. **Suslow, T.** 1997. *Postharvest Chlorination: Basic Properties and Key Points for Effective Disinfection.* Publication 8003. [Online at http://ucce.ucdavis.edu/files/datastore/234-404.pdf.] Division of Agriculture and Natural Resources, University of California, Davis, Calif. http://danrcs.ucdavis.edu.

48. **Tauxe, R. V.** 1997. Emerging foodborne diseases: an evolving public health challenge. *Emerg. Infect. Dis.* **3**:425–434.

49. **Thurston-Enriquez, J. A., P. Watt, S. E. Dowd, R. Enriquez, I. L. Pepper, and C. P. Gerba.** 2002. Detection of protozoan parasites and microsporidia in irrigation waters used for crop production. *J. Food Prot.* **65**:378–382.

50. **Turner, C.** 2002. The thermal inactivation of *E. coli* in straw and pig manure. *Bioresource Technol.* **84**:57–61.

51. **Vlahovich, K. N., E. A. Bihn, R. B. Gravani, R. W. Worobo, and J. J. Churey.** 2004. The detection and survival of *Salmonella, Escherichia coli,* and *Listeria monocytogenes* in selected pesticide sprays for use on fresh produce, p. 120. P212. *Int. Assoc. Food Prot. 91st Annu. Meet. Program Abstr. Book.* International Association for Food Protection, Phoenix, Ariz.

52. **Warrington, P. D.** 8 March 1998, posting date. Water Quality Criteria for Microbiological Indicators: Overview Report. [Online.] Ministry of Water, Land and Air Protection, Resource Quality Section, Water Management Branch, Ministry of Environment and Parks, British Columbia, Canada. http://wlapwww.gov.bc.ca/wat/wq/BCguidelines/microbiology.html.

Microbiology of Fresh Produce
Edited by Karl R. Matthews

Biology of Foodborne Pathogens on Produce

3

Ethan B. Solomon, Maria T. Brandl, and Robert E. Mandrell

In the past decade, the number of outbreaks of foodborne illness arising from the consumption of fresh fruits and vegetables has increased dramatically (8). Numerous reasons have been put forth to explain this increase, including changes in dietary habits, modifications in agronomic and processing practices, increased international trade and distribution, and advances in epidemiological surveillance and microbiological detection methods (8, 15). *Escherichia coli* O157:H7 and *Salmonella enterica*, both previously regarded as pathogens linked to foods of animal origin, have emerged as the two most common causative agents of produce-related outbreaks (20).

Microbiological research initially focused on surveys of fresh fruits and vegetables to determine the prevalence of pathogens (3, 4), coupled with an intensive effort to find interventions capable of removing or inactivating bacteria attached to postharvest produce (2, 43, 67, 86). Results of these efforts demonstrated that while the incidence of pathogens was low, none of the sanitizers, rinses, or treatments proposed was capable of producing pathogen-free produce. A complete discussion on postharvest sanitizing strategies is presented in chapter 4.

The continued rise in the number of produce-related outbreaks and the realization that no sanitizer was completely effective have led to intensive research into the ecology of human pathogens in the plant environment. Bacterial human pathogens have been shown to survive for long periods in

ETHAN B. SOLOMON, DuPont Chemical Solutions Enterprise, Experimental Station Laboratory—E402/3223, P.O. Box 80402, Wilmington, DE 19880-0402. MARIA T. BRANDL AND ROBERT E. MANDRELL, Produce Safety and Microbiology Research Unit, Western Regional Research Center, Agricultural Research Service, U.S. Department of Agriculture, Albany, CA 94710.

water, animal manures, and soils, as well as on growing plant tissue (33, 34, 60, 113, 114). More recently, the intimate interactions between human pathogenic bacteria and plants have begun to be scrutinized. A limited number of publications have demonstrated that human pathogens can attach to and colonize the tissues of growing plants (11, 23).

The goal of this chapter is to review the attachment to and localization of human pathogens on the surfaces of plants. We first describe the physicochemical environment of plant surfaces to which bacteria must attach to survive and proliferate and then summarize the major known mechanisms of attachment of bacteria to plants, with special reference to the similarities between the cell surface moieties of plant-associated bacteria and those of human pathogens. We also discuss the research that has been conducted on the localization of human pathogenic bacteria on plant surfaces. Finally, we present experimental evidence for the internalization of human pathogens in plants, as well as for their persistence on and in field-grown crops.

THE PLANT ENVIRONMENT

Plant surfaces are colonized by a variety of bacteria, some of which are pathogens of plants and have the ability to invade plant tissue. Bacterial cells that land on plant surfaces in the field encounter a harsh environment with large and rapid fluctuations in temperature, humidity, osmotic potential, and water and nutrient availability (44, 49). It is the ability of immigrant foodborne pathogenic bacteria to attach and grow and/or survive in this environment that determines their fate on contaminated fruits and vegetables.

Aerial Surfaces

The surface of aerial plant parts, such as leaves and fruits, is called the cuticle. The cuticle is a matrix of variable thickness composed of cutin (a biopolyester composed mainly of fatty acids), waxes, and polysaccharides (59). The cuticle acts as a barrier to protect epidermal cells and prevent water loss from the plant (59), and it is covered frequently with a waxy layer with three-dimensional crystalline structures (76). The cuticular waxes are lipophilic long-chain fatty acids, some of which are oxygenated, thus forming aldehydes, ketones, sterols, and esters (59).

At the scale of bacterial cells, the topography of aerial plant surfaces is complex. This complexity is caused not only by the presence of wax crystals but also by that of plant hairs (trichomes) in some plant species and breaks in the cuticle that can form crevasses. In addition, the epidermal cells and their junctions produce bulges and depressions. This waxy landscape, also called the phylloplane, is highly hydrophobic and partly governs the attachment and the interactions of immigrant bacteria with the plant surface (89). Breaks in the cuticle, or lesions that expose epidermal cells, offer additional sites for microbial attachment to or interaction with the plant tissue.

Quantification and characterization of leaf exudates revealed that nutrients on leaves are composed primarily of sugars, amino acids, and organic acids (28). There is evidence that the growth of bacteria on leaves is limited by carbon sources rather than nitrogen (119). Simple carbohydrates, such as glucose, fructose, and sucrose, are the most common carbon sources available to bacteria on leaves (77). Studies with whole-cell bacterial biosensors have shown that the spatial distribution of sucrose (78) and fructose and glucose (65) on leaf surfaces is heterogeneous. Thus, oases of abundant sugars exist while most of the leaf surface is relatively oligotrophic (65). The availability of Fe^{3+} also is highly variable on the phylloplane (61). Other aspects of the physicochemical environment of plant surfaces, such as UV radiation and water availability, are important factors that shape bacterial colonization of leaves and have been reviewed previously (68).

Root Surfaces

The rhizoplane is defined as the surface of roots together with soil particles closely adhering to it (22). Root surfaces are immersed in mucigel, a substance composed of plant-secreted mucilage (composed of pectins and hemicelluloses) and complex polysaccharides produced by bacteria that also degrade mucilage (25, 26). Depending on the plant, root exudates contain a variety of substances that can act as chemoattractants for microorganisms and/or as substrates for growth. Sugars, amino acids and other amino compounds, organic acids, fatty acids and sterols, growth factors, nucleotides, and other compounds are released from root cells (25). Therefore, the surface of roots can be a zone of intense microbial activity. In particular, nutrients exuded at the junction of lateral roots and at the root tip may form energy-rich microenvironments where bacteria can thrive (58, 112). The rhizoplane is characterized by rapid changes in physicochemical conditions brought about by rain and/or irrigation, drought, and plant transpiration and by continuous changes due to root growth and interactions with soil and root microflora (10, 44).

ATTACHMENT OF BACTERIA TO PLANT SURFACES

The initial step resulting in the contamination of fresh produce is the attachment of human pathogenic bacteria to plant tissue. Recent studies of *E. coli* O157:H7 on store-bought lettuce indicated that cells attach in a relatively short time period and that not all cells can be removed by vigorous washing or treatment with chlorine (66, 92, 103, 109). Similar observations have been reported for other enteric pathogens and plant species (84, 106). These observations provided the first evidence that enteric pathogens have the ability to attach to plant tissue. More important, differences in the levels of attachment of various enteric pathogens to plants suggested that specific mechanisms of attachment were involved in this interaction. For example, *S. enterica* and *E. coli* O157:H7 strains attached in higher numbers than *Listeria monocytogenes* strains to the surfaces of whole cantaloupes (107). In a separate study, a comparison of multiple *E. coli* O157:H7 and *S. enterica* strains on alfalfa sprouts revealed that *E. coli* O157 attached to sprout tissues at a level 10- to 1,000-fold lower than *S. enterica.* In addition, four nonpathogenic *E. coli* strains isolated from field-grown cabbage attached to sprouts better than *E. coli* O157:H7 but at a lower level than *S. enterica* serovar Newport (6). In contrast, in a lettuce sprout model system, strains of *E. coli* O157:H7 implicated in produce outbreaks attached to lettuce roots approximately 10-fold better than did two of five nonpathogenic *E. coli* strains (110). Such differences in attachment profiles have been observed also with strains of *L. monocytogenes*. While minimal differences among levels of attachment of seven strains of *L. monocytogenes* to cut radish tissue were detected (38), 100- to 1,000-fold differences were reported for levels of attachment to and colonization of alfalfa sprouts by different *L. monocytogenes* strains (39). Thus, there appears to be high variability in attachment abilities at the strain level, as well as among plant model systems. Explanations for these differences may lie in several mechanisms of bacterial attachment described below.

Mechanisms of Attachment

Most of the fundamental and definitive studies of plant-microbe interactions, including attachment, involve gram-negative plant-symbiotic or pathogenic bacteria. For example, *Rhizobium* species, common endosymbionts involved in nodule formation and nitrogen fixation on legume roots, provide the most advanced model of microbial attachment to, and interactions with, plants.

It is probable that some of the human pathogens that contaminate crop plants use mechanisms of attachment similar to those employed by plant-associated bacteria. Thus, any similarities in the localization patterns of human pathogens and plant-associated bacteria on plants, in the biochemistry profiles of bacterial outer surface structures, and in adhesins may serve as the basis for fundamental studies on the attachment of human pathogens to plants. In this section, we will examine some of the best-described factors of bacterial attachment to plants and provide a context for assessing observations obtained from studies of human pathogens in plant environments. Table 1 lists some of the well-described plant attachment factors in plant-associated and foodborne pathogenic bacteria. Figure 1 provides a schematic of selected adhesins.

Nonfibrillar proteinaceous adhesins

Although several adhesins that have a role in bacterial attachment to animal or plant hosts are pili or fimbriae, some are nonstructural and consist of small proteins that are exposed on the outer surface of the bacterial cell. For example, the first step in the interaction of *Rhizobium* spp. with leguminous plants involves a bacterial Ca^{2+}-binding protein, rhicadhesin, which is responsible for the attachment of mostly single (not aggregated) bacterial cells directly to the root hair (94). Rhicadhesin was reported to be involved also in the attachment of *Agrobacterium tumefaciens,* a plant pathogen that produces galls on stems and roots, to pea root hair tips (101). Application of

Table 1 Major plant attachment factors described in bacteria

Attachment factors	Bacterial species	Reference(s)
Afibrillar adhesins	*R. leguminosarum*	94
	A. tumefaciens	101
	Azospirillum brasilense	16
	Erwinia chrysanthemi	87
Cellulose fibrils	*R. leguminosarum*	93
	A. tumefaciens	74
Polysaccharides	*Azospirillum brasilense*	27
	Azospirillum lipoferum	27
	A. tumefaciens	73, 75, 118
	E. coli O157:H7[a]	45
Bacterial lectins	*B. japonicum*	50, 69
	R. leguminosarum	5
	Ralstonia solanacearum	99, 100
Fimbriae		
Type I	*K. pneumoniae*	42
	Enterobacter agglomerans	42
Type III	*K. mobilis*	63
Type IV	*X. vesicatoria*	81
Pili	*P. syringae* pv. *phaseolicola*	48, 49, 90
	P. fluorescens	108
	A. tumefaciens	37
Flagella	*Azospirillum brasilense*	24
	L. monocytogenes	38

[a]Putative attachment factor.

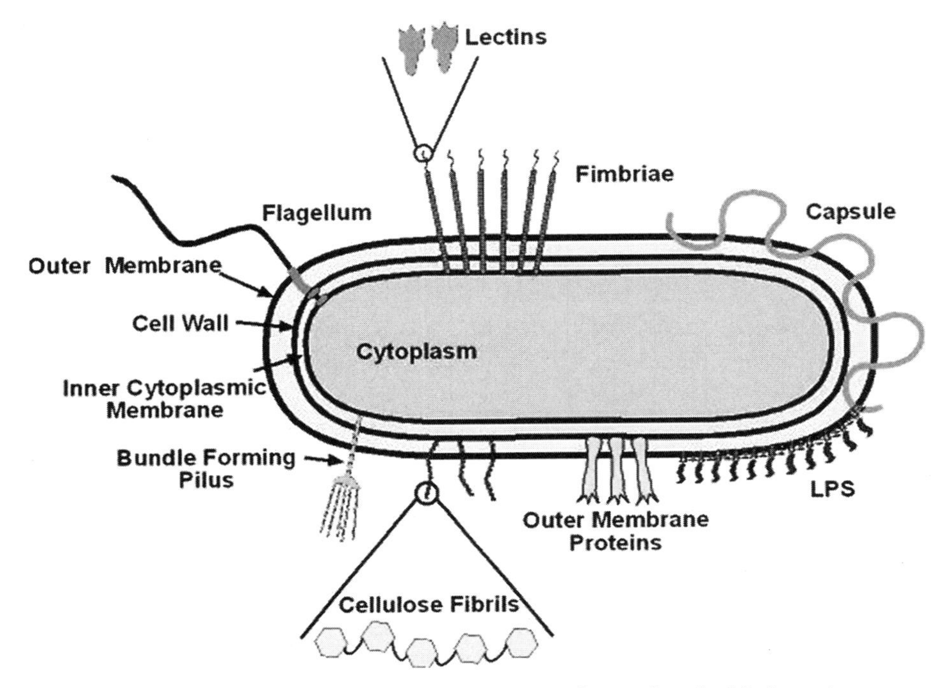

Figure 1 Schematic representation of select attachment factors described in bacteria. (Artwork courtesy of Nereus Gunther, U.S. Department of Agriculture, Agricultural Research Service, Eastern Regional Research Center, Wyndmoor, Pa.)

rhicadhesin inhibits the attachment of many *Rhizobiaceae* species to pea root hair tips, including *Rhizobium leguminosarum* biovars, *Rhizobium meliloti*, *Rhizobium lupini*, and *Bradyrhizobium japonicum,* as well as *A. tumefaciens* and *Agrobacterium rhizogenes*, indicating that rhicadhesin or rhicadhesin homologs are part of a common mechanism of attachment to root hairs (94). This would apply also to *Azospirillum brasilense*, in which the major outer membrane protein was reported to be an adhesin involved in the variability in the levels of attachment of *Azospirillum brasilense* to root extracts of various plant species and in bacterial cell-cell aggregation (16).

The soft-rot pathogen *Erwinia chrysanthemi* produces an adhesin (HecA) that contributes to its virulence, aggregation on, and attachment to tobacco (87). The homology of *hecA* to genes encoding hemagglutinins in both plant and animal pathogens suggests the possibility that nonfibrillar adhesins in foodborne pathogenic bacteria may mediate attachment not solely to animal cells but also to plant cells.

Cellulose fibrils

In *A. tumefaciens* and *R. leguminosarum*, the second step in attachment involves the synthesis by the bacteria of cellulose fibrils that bind the bacteria tightly to the plant host surface and lead to autoaggregation and/or firm binding of other bacteria at the site of infection (74, 93). In *E. coli* and *S. enterica*, cellulose biosynthesis is involved in multicellular behavior and biofilm formation in vitro (122), but its role in adhesion to plant cells has not been demonstrated. Also, several additional species of the *Enterobacteriaceae*, including some that are commonly isolated from plants, such as *Citrobacter* spp. and *Enterobacter* spp., have the ability to produce cellulose (121). Gal et al. (35) reported that production of a cellulosic polymer in *Pseudomonas fluorescens* contributed to the fitness of this common colonizer of plants in the phyllosphere and the rhizosphere. It remains to be determined whether cellulose has a function in the attachment to plants by most enteric or plant-associated bacteria that produce it (121).

Polysaccharides

Exopolysaccharides and lipopolysaccharides (LPS), or lipooligosaccharides, are prominent cell surface glycoconjugates in gram-negative bacteria and potential attachment factors as receptors for animal (71) or plant lectins. Two different polysaccharide structures were identified in *Azospirillum brasilense* and *Azospirillum lipoferum:* a capsular polysaccharide (CPS) tightly associated with the cell surface and an exopolysaccharide appearing to be less dense and extending from the cell. Wheat germ agglutinin, a lectin, bound to cells from both species; this binding was inhibited by *N*-acetylglucosamine, providing evidence of candidate carbohydrates for binding to plant lectins (27). Earlier studies reported that EPS is the probable bacterial receptor responsible for specific interactions between *B. japonicum* cells and soybean root hairs by means of a soybean lectin (9, 105).

Whatley et al. (118) demonstrated that both LPS on *A. tumefaciens* cells and purified LPS inhibited specifically the tumorigenic activity of *A. tumefaciens* on pinto bean leaves by >50%. Additionally, LPS purified from the parent and attachment-impaired mutant strains inhibited the attachment of *A. tumefaciens* to carrot cells by 30 to 60%. These observations suggested that LPS interacts with the sites of attachment on the plant (73). In other studies, a polysaccharide extracted from *A. tumefaciens* strain C58 inhibited the strain's attachment to carrot cells (85), and a mutant (AttR$^-$) of this strain lacked an acetylated CPS and did not attach to wound sites on plants (75). Interestingly, the ligand in carrot cells that bound the polysaccharide may be a homolog of vitronectin (S protein), a serum-spreading factor in animals and part of the extracellular matrix (111). The vitronectin-like protein was detected on carrot cells (111) and on tomatoes, soybeans, and broad beans (91). The attachment of *A. tumefaciens* via a polysaccharide to an animal protein homolog found on a variety of plants provides another example of the potential for common strategies of attachment of plant and foodborne pathogens to plant tissue.

It was reported that the medium in which *E. coli* O157:H7 is grown affects attachment to plant tissue. Cells grown in tryptic soy broth were more hydrophilic, produced more CPS, and attached better to edges of lettuce (0.4 $\log_{10}$) and to the surfaces of both lettuce and apples (0.8 to 1.0 $\log_{10}$) than those grown in nutrient broth (45). This suggested that CPS may be involved directly in attachment of *E. coli* O157:H7 to plants. Further studies are needed to confirm this hypothesis.

Bacterial lectins

In addition to harboring carbohydrate receptors for attachment via plant lectins, there is evidence that bacteria produce lectins that may mediate their interaction with carbohydrates present on plant surfaces. In *B. japonicum*, a cell surface lectin involved in adhesion to roots was identified and shown to be inhibited by lactose and galactose (50, 69). A lectin similar to that in *B. japonicum*, RapA1, was described for *R. leguminosarum*. Both lectins are unipolarly located on the cells and have activity in agglutination of bacteria (5). It is thought that these bacterial lectins are capable of recognizing carbohydrates on bacterial cells and on root hairs and thus serve in aggregation as well as attachment to plants. Recently, Sudakevitz et al. (99, 100) have characterized two lectins in *Ralstonia solanacearum* that have high affinity for sugars present in or on plants. It is noteworthy that one of the lectins, RS-IIL, is similar to the PA-IIL lectin of *Pseudomonas aeruginosa*, which contributes to the aggregation, host specificity, and virulence of the bacterium (100). The similarity of a plant pathogen lectin to a human pathogen lectin is one more parallel worthy of exploring to decipher the mechanisms of attachment of bacteria, including enteric pathogens, to plant surfaces. For example, the preferential attachment of *L. monocytogenes* and *E. coli* O157:H7 to the cut edges of lettuce leaves (103) may be related to the interaction of lectins on the surfaces of these pathogens with specific carbohydrates that leak from the damaged plant tissue.

Structural adhesins

Foodborne pathogenic bacteria produce structural adhesins that play a role in their attachment to host cells and/or in their pathogenicity and virulence. These structural adhesins, known as fimbriae (pili) and flagella, are the best-characterized attachment factors in enteric bacteria. Structural adhesins, which are remarkably conserved among major groups of bacteria, have a distinct role also in the interaction of plant-associated bacteria with their hosts. The following section describes some of these adhesins and their function in plant-microbe interactions. It remains unclear at this point whether any of these adhesins are important in the attachment of enteric pathogens to plants. However, the observation that bacteria such as *Klebsiella* spp., *Enterobacter* spp., and *Pseudomonas* spp. can associate with both plants and animals may lead to the discovery of attachment factors that enable enteric bacteria to bridge these environments.

Fimbriae and pili

Fimbriae (pili) are proteinaceous hair-like appendages that are common in gram-negative bacteria. *Klebsiella* spp. are enteric bacteria that can be soil borne and saprophytic and cause serious human illness. In a nitrogen-fixing strain of *Klebsiella mobilis* (renamed *Enterobacter aerogenes*) isolated from plants, type III fimbriae were associated with hemagglutination of human O erythrocytes and adhesion to *Poa pratensis* seedling roots (63). In addition, the purified fimbriae bound directly to root tissue. Subsequently, type I fimbriae were reported to mediate adherence of *Klebsiella pneumoniae* and *Enterobacter agglomerans* to roots; this binding was inhibited by α-methyl-D-mannoside (42).

Type IV bundle-forming fimbriae, composed mainly of FimA protein subunits, have been characterized for the common plant pathogen *Xanthomonas vesicatoria* (81). Although a $FimA^-$ mutant strain was not impaired in adhesion to tomato leaves, the parental strain was more prevalent and aggregated on leaf trichomes and in infected leaves, suggesting that type IV fimbriae assist in the attachment or aggregation of the pathogen at specific locations on plants (81).

Pseudomonas spp. are common and successful colonizers of plant surfaces. *Pseudomonas syringae* pv. *phaseolicola*, which causes disease on bean plants, attaches to the stomata of bean leaves via pili; a pilus-negative mutant of this pathogen was washed more easily from leaves than the parental strain (48, 49, 90). Also, piliated or fimbriated strains of *P. fluorescens*, a common nonpathogenic colonizer of roots, were shown to attach better than a nonpiliated variant strain to roots of corn seedlings (108). In *A. tumefaciens*, the conjugative pilus is required for plant transformation and may be used also for attachment of the pathogen to the plant, since *Arabidopsis thaliana* plant lines have been identified that resist both *A. tumefaciens* transformation and attachment to root hairs (37).

Flagella

The role of flagella in the attachment of plant-associated bacteria has not been fully explored because of the primary effect of flagellar mutations on bacterial motility and, thus, the potential for flagellar mutants to generate biased data in attachment assays. However, evidence for the involvement of flagella in bacterial attachment to plants was demonstrated with *Azospirillum brasilense.* A nonmotile flagellar mutant of this species, which possesses a polar flagellum, was impaired dramatically in attachment to wheat roots. Additionally, the purified flagella from the parental strain bound directly to wheat roots (24).

In what is perhaps, so far, the best attempt to decipher the molecular factors that underlie the attachment of an enteric pathogen to plants, Gorski et al. (38) screened a mutant library of *L. monocytogenes* for attachment to radish slices. Three mutants were identified that were reduced by at least 10-fold in attachment to radish tissue compared to the parental strain when incubated at 30°C. Two mutations were in genes of unknown function within an operon encoding flagellar biosynthesis, but only one of the corresponding mutants was nonmotile. The third mutant carried an insertion in a gene necessary for the transport of arabitol. The motility mutant was impaired in attachment also at 10 and 20°C, whereas the arabitol transport mutant showed reduced attachment also at 10°C. Interestingly, no difference in attachment levels between the mutants and the parental strain was observed when the assay was performed at 37°C. The effect of temperature on attachment in this system suggests that *L. monocytogenes* may express different plant attachment factors under different environmental conditions.

With the use of antibodies specific to the flagella of *S. enterica* serovar Thompson, this pathogen was detected by immunoelectron microscopy after inoculation onto and incubation on cilantro plants (12). More important, the high resolution of the electron microscope allowed for observation of the flagella which appeared to anchor serovar Thompson cells to the leaf surface, suggesting that they may serve for attachment to plants. Although electron microscopy has been used frequently to visualize structural adhesins of pathogenic bacteria grown in culture, this powerful tool has been applied rarely to the investigation of the structural attachment factors of bacteria in their habitats.

LOCALIZATION

Research using novel microscopy techniques such as green fluorescent protein (GFP) labeling and confocal microscopy has provided evidence that bacteria attach to and colonize plant tissue. The following section presents an overview of the many observations regarding the localization of foodborne pathogenic bacteria after their inoculation onto whole plants or plant tissue under laboratory conditions.

Leaves

The investigation of outbreaks of *S. enterica* (19) and *E. coli* O157:H7 (47) infection, linked to cilantro and lettuce, respectively, suggested that the outbreaks resulted from preharvest contamination of the produce. Inoculation of cilantro plants with the GFP-labeled outbreak strain of *S. enterica* serovar Thompson and incubation under warm and humid conditions revealed that this pathogen formed microcolonies preferentially in the vein area of the leaf (11). Also, high densities of serovar Thompson cells were found on the veins of senescent leaves after prolonged incubation times (Fig. 2A). The pathogen was located at a higher frequency on the veins than on other areas of the leaf at the time of inoculation also, although only individual cells of serovar Thompson, scattered at distant locations, were present. Thus, there was a correlation between the attachment of the serovar Thompson inoculum cells and the presence of aggregates following colonization. Higher bacterial incidence in the vein area was reported also for *E. coli* O157:H7 on lettuce leaves (92) and for epiphytic bacteria on various plant species (64, 65). The enhanced wettability of this part of the leaf (64) may allow for easier contact of inoculum cells with the plant or enhanced attachment. Also, it is possible that specific factors on the surfaces of leaf veins, where nutrients are more abundant (65), facilitate the attachment of bacteria. Investigation of the attachment of *S. enterica* to polymers purified from cilantro leaf surface extracts demonstrated that the pathogen binds strongly to stigmasterol (72), a compound detected also in other leaf extracts (31). An investigation of the distribution of stigmasterol may reveal whether this compound is present evenly on the leaf surface or localized in specific regions of the leaf.

In a recent study, cut lettuce leaves immersed in a suspension of *E. coli* O157:H7 were exposed to fluorescent antibody specific to this pathogen and observed by confocal microscopy. *E. coli* O157:H7 attached predominantly to the cut edges of leaves; fewer cells attached to the intact cuticle region of leaves but were observed near stomata, on trichomes, and on veins (92). Strains of *E. coli* O157:H7, *P. fluorescens*, *S. enterica*, and *L. monocytogenes* attached to different regions of cut lettuce leaves, indicating differences in mechanisms or strengths of attachment among bacterial species (103). Hassan and Frank (46) addressed the general nature and force of the lettuce leaf-*E. coli* O157:H7 interactions by measuring the effects of surfactants and

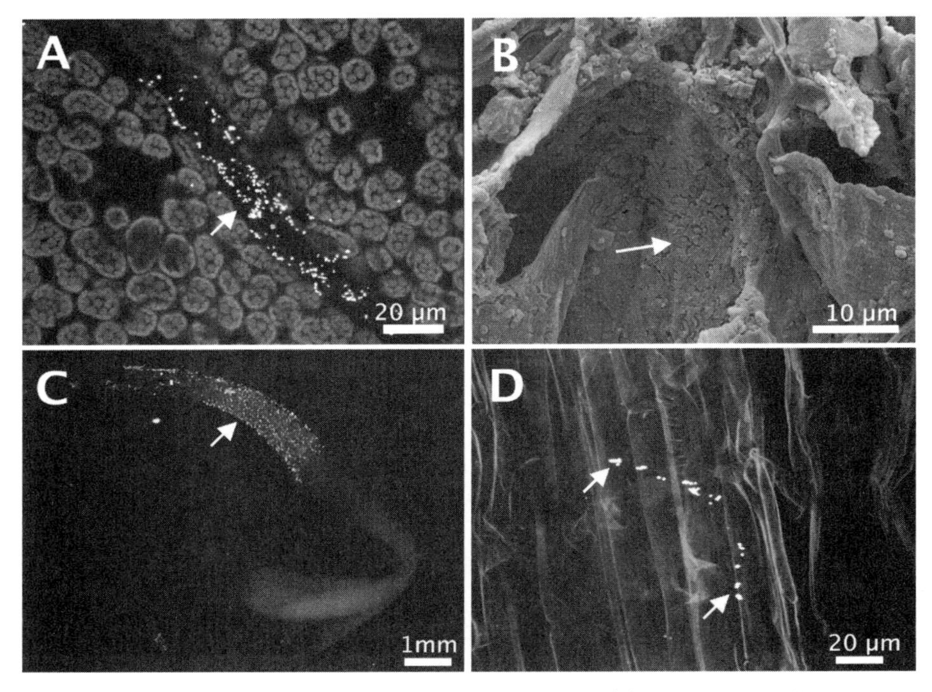

Figure 2 Fluorescence (A, C, and D) and scanning electron (B) micrographs of foodborne pathogenic bacteria (arrows) that attached to and colonized various plant surfaces after their inoculation in the laboratory. (A) GFP-labeled *S. enterica* serovar Thompson on the vein of a cilantro leaf (micrograph by M. T. Brandl). (B) *S. enterica* on the rind surface of a cantaloupe (micrograph by E. B. Solomon). (C) GFP-labeled *S. enterica* serovar Newport on root hairs of alfalfa sprouts (21). (D) GFP-labeled *E. coli* O157:H7 on the epidermal cells of lettuce roots irrigated with contaminated water (110).

various solutions on attachment. The most hydrophobic surfactants were the most effective in detaching the pathogen from the leaf cuticle, but cells at cut edges remained attached. The level of bacterial attachment to the lettuce leaf surface was sixfold higher after treatment with $CaCl_2$, but a minimal difference was observed on the cut edges of leaves, and treatment with NaCl also had a minimal effect (46). These studies suggest that hydrophobic interactions, neutralization of ionic charge, or bridging of anionic moieties by divalent cations is possibly involved in the attachment of *E. coli* O157:H7 to different leaf tissues.

Fruits

Cantaloupe melons and tomatoes have been associated with several multistate outbreaks of salmonellosis in the United States. Ukuku and Fett (107) reported the most in-depth investigation of the type and strength of attachment of multiple strains of *S. enterica*, *E. coli* (O157:H7 and non-O157:H7), and *L. monocytogenes*, including outbreak- and food-associated strains, to the surfaces of cantaloupes. The strength of the interaction between each pathogen and the fruit surface was measured based on the number of bacterial cells retained on the cantaloupe surface after immersion of whole melons in water and was then considered in the context of the bacterial cell surface charge and the hydrophobicity of each strain. *S. enterica* had the highest and most variable surface hydrophobicity and the highest negative and positive surface charges. Although more *E. coli* cells attached initially to the melon surface than cells of the other strains, *S. enterica* cells bound to the fruit surface more strongly than either *E. coli* or *L. monocytogenes* cells at 4°C on days 0, 3, and 7 after inoculation. The strength of attachment of each species correlated directly with the hydrophobicities and the surface charges of the strains, indicating a linear relationship between these parameters and attachment to cantaloupe rinds, which are highly hydrophobic (107). Figure 2B shows a scanning electron micrograph of *S. enterica* following inoculation (10^3 CFU) onto and 24-h incubation on the outer surface of a cantaloupe.

In a separate study, *L. monocytogenes* attached to unwaxed whole cucumbers better than to waxed cucumbers, indicating that the hydrophilic nature of the *L. monocytogenes* cell surface has a role in attachment in that model (84). This is consistent also with the observation that this pathogen attached in higher numbers to the cut surfaces of lettuce leaves than to their undamaged and, thus, more hydrophobic adaxial and abaxial surfaces (103).

Although the attachment of *S. enterica* to tomato fruits has not been investigated specifically, several studies have demonstrated that an outbreak strain of *S. enterica* serovar Montevideo has the ability to survive for prolonged periods of time on the surfaces of intact fruits, with increased populations in cracks and in stem scars (40, 117). These observations suggest that in addition to the better opportunity for growth in the damaged tissue of tomatoes, the strength of the attachment of serovar Montevideo is possibly higher in these sites, as shown in other studies described above.

Apples provide a surface and environment for bacteria similar to those of tomatoes. The intact skin of apples is composed of a waxy cuticle less conducive to attachment by human pathogens than other regions of the fruit. Liao and Sapers (67) reported that after inoculation onto whole apples, 94% of *S. enterica* serovar Chester cells were located on the stems and calices and only 4% were located on the skin of the remaining area of the fruits.

The recurrent observation that human pathogens localize in higher numbers on broken plant tissue and on the less waxy plant surfaces is difficult to dissociate from the fact that higher volumes of aqueous inoculum suspensions would be deposited on less hydrophobic areas during inoculation. Thus, the hypothesis that human pathogens attach better or with higher strength, per se, to broken or less waxy plant tissue still remains to be tested fully.

Roots

Numerous outbreaks of enteric illness have been caused by the contamination of sprouts with *E. coli* O157:H7 and *S. enterica* (13, 70, 79). At least one of these outbreaks was linked to contaminated seeds (13). Sprouts grown from seeds contaminated with human pathogens in vitro in the laboratory have served as a model to study the attachment of enteric pathogens to roots. Several studies have reported the attachment and growth of *E. coli* O157:H7 (110), *S. enterica* (21), and *L. monocytogenes* (39) on the broken seed coat edge of germinating seeds inoculated with these human pathogens. In addition, the above-mentioned studies demonstrated that all three pathogens adhere to and grow preferentially in the root hair zone. Thus, adhesion of human pathogens to roots under these laboratory conditions followed localization patterns similar to those reported earlier for various plant-associated bacteria (88). The observation that *S. enterica* formed large aggregates on alfalfa root hairs (Fig. 2C) compared to *E. coli* O157:H7, which was present mainly as single cells or small aggregates (21), suggests that a second step in attachment that also leads to autoaggregation may be present in *S. enterica*. This second step is mediated by the production of cellulose fibrils in *A. tumefaciens* and *R. leguminosarum* (74, 93), a pathway described also for *S. enterica* (122) but not associated yet with its attachment or aggregation in the rhizosphere.

Few studies have attempted to investigate the localization of human pathogens on the roots of plants grown in soil. Cooley et al. (23) observed that, at early incubation times, *E. coli* O157:H7 and *S. enterica* cells were concentrated at the root tips and branch points of lateral roots of *Arabidopsis thaliana* plants grown in soil inoculated in the laboratory. At later incubation times, the roots were colonized more uniformly by the pathogens. It remains unclear whether the temporal difference in colonization patterns was caused by differences in attachment levels or in localized growth patterns of these enteric bacteria. In a separate study in which lettuce seedlings grown in soil were irrigated with a suspension of *E. coli* O157:H7, the pathogen was detected on various parts of the root, including epidermal cells (Fig. 2D) (110). No evidence was found that the pathogen attached to or grew in a localized fashion on the roots. The above-described studies underscore the great variability in results among different experimental plant model systems and the fact that general conclusions regarding the behavior of human pathogens on plant surfaces should be drawn with caution.

INTERNALIZATION

Antimicrobial agents such as chlorine, hydrogen peroxide, ozone, trisodium phosphate, and peroxyacetic acid all have been examined as postharvest sanitizers for use on fresh produce (2, 17, 86). None has been demonstrated to be completely effective in eliminating inoculated bacteria. Recent research has demonstrated the ability of enteric pathogens to become internalized within a number of fruits and vegetables (115). This ability to infiltrate plant tissues is of grave concern in the food industry since microorganisms present within plant structures are protected from surface decontamination.

Internalization can occur through natural openings such as stem scars, stomata, lenticels, and broken trichomes (51, 83). Damage to the waxy cuticle on plant tissues also has been shown to provide a point of ingress for bacterial pathogens (62). Last, there has been considerable interest in the ability of both *S. enterica* and *E. coli* to exist as endophytes within growing plants via colonization of the root system.

Internalization through Natural Openings

Enteric bacteria penetrate into a wide variety of commodities through natural openings. Apples have been demonstrated to be susceptible to internalization following immersion in bacterial suspensions. The blossom ends of intact apples allow uptake of red dye as well as *E. coli* O157:H7 (14). Burnett et al. (18) utilized confocal microscopy to demonstrate infiltration by GFP-expressing *E. coli* O157:H7 through the floral tubes of apples following immersion of apples in cell suspensions. Bacteria could be found adhering to trichomes, crevices, and ridges within the ventral cavities and seed locules of intact apples. In addition, external structures such as intact skin and lenticels allowed for penetration to approximately 30 μm beneath the surface. Intact oranges have been demonstrated to be susceptible to infiltration by *S. enterica* and *E. coli* O157:H7 through the stem scar (30). The above-described studies revealed that internalization was enhanced when experiments were carried out under negative temperature differentials (warm fruit and cold inoculum). Other fruits such as mangos (82) and tomatoes (7, 120) also are susceptible to bacterial infiltration through the stem scar. The results of these studies emphasize the difficulties in decontaminating fresh fruits intended for juice production.

Stomata have been identified as natural openings that permit the entry of bacterial cells into both lettuce (92, 102) and radish sprouts (57). *E. coli* O157:H7 inoculated onto cut lettuce pieces was found entrapped within stomata approximately 20 μm beneath the leaf surface (92). Bacterial cells localized within stomata were protected from sanitation with chlorine (102).

E. coli O157:H7 was detected within the inner tissues of radish sprouts grown from contaminated seeds (57). Scanning electron microscopy indicated that the stomata of radish sprouts were colonized by large numbers of bacteria.

Internalization through Damaged Tissue

The internalization of foodborne pathogens through damaged plant tissues has been established for a wide range of commodities. Using cut lettuce leaves as a model system, Seo and Frank (92) visualized antibody tagged-*E. coli* O157:H7 at locations 20 to 100 μm below the cut surface. Additional work by Takeuchi and Frank (102) indicated that cells attached at subsurface locations were protected from sanitation with chlorine at a concentration of 200 ppm. Studies with apples demonstrated penetration by *E. coli* O157:H7 to 70 μm below the skin of the fruits in sites surrounding puncture wounds (18). In addition, apples bruised by being dropped from a height of 1.5 m were more likely to harbor cells of *E. coli* O157:H7 below the surface than unbruised fruits (62). Analysis of confocal laser scanning micrographs indicated that cells could be detected at depths of up to 30 μm below the surfaces of inoculated, unwashed apples. Cells remaining near the surface were primarily entrapped within waxy platelets and not removed by washing. Han et al. (43) observed infiltration of green peppers by *E. coli* O157:H7 after application of cells onto coarse, porous, or injured surfaces of the fruits. The authors speculated that waxes on the surfaces of uninjured peppers provided a barrier against infiltration by *E. coli* O157:H7 but that disruptions in the waxy cuticle allowed for attachment and penetration by a large number of bacteria.

Preharvest Internalization

Given that the incidence of outbreaks arising from fresh produce has increased significantly only in the past decade, studies of the survival of enteric bacteria in the field or in soil microcosms were initiated only recently (32-34, 80, 114). The survival of *E. coli* and *S. enterica* for extended periods of time in water, manure, and soil suggested the possibility that plants could become contaminated in a preharvest fashion in the field. Since then, a number of studies have focused on the ability of human pathogens to colonize and become internalized in crop plants such as tomatoes, lettuce, sprouts, and spinach.

Guo et al. (40, 41) performed a series of experiments on the interactions between *S. enterica* and growing tomato plants. In the first study, *S. enterica* cells were inoculated onto tomato plants before and after flower set by injecting bacteria into stems or brushing flowers with the inoculum (40). Plants were then allowed to bear fruit, and ripe tomatoes were picked and subjected to microbiological analysis. Results indicated that 11 of 30 tomatoes harvested from inoculated plants harbored the pathogen. The pulp of 6 of the 11 positive tomatoes contained *S. enterica,* suggesting systemic movement of the organism from the site of stem inoculation. A second study focused on the internalization of bacteria in hydroponically grown tomato plants (41). Following 1 day of exposure to hydroponic solution inoculated with *S. enterica*, the hypocotyls, cotyledons, and stems of tomato seedlings contained the human pathogen. After 9 days of exposure, the leaves, stems, hypocotyls, and cotyledons of plants with both intact and injured root systems contained high levels of *S. enterica.* These results demonstrated the uptake and internalization of this pathogen, providing further evidence for the ability of enteric bacteria to be transported systemically within growing plants.

The association between *E. coli* O157:H7 and lettuce has been investigated by a number of laboratories both because of the plant's commercial significance and because of repeated outbreaks linked to lettuce consumption (1, 47). Wachtel et al. (110) studied the interactions between *E. coli* O157:H7 and lettuce seedlings in both hydroponic and soil growth conditions. Results indicated that the roots, hypocotyls, and cotyledons of soil-grown plants inoculated with contaminated irrigation water harbored the pathogen for 10 days, regardless of the level of bacteria applied to plants (10^2 to 10^8 log CFU/ml). Examination of seedlings by confocal microscopy revealed that *E. coli* O157:H7 cells adhered to guard cells surrounding stomata and moved within the vasculature of intact hypocotyls. The authors speculated that these bacteria were moving through the xylems, as they are the only open vessels within the seedlings. The transmission of *E. coli* O157:H7 from contaminated soil and irrigation water was also reported by Solomon et al. (97). Seedlings were grown in a manure-soil mixture containing *E. coli* O157:H7. A 10-min exposure of the exterior of harvested seedlings to $HgCl_2$ failed to eliminate all culturable cells of the pathogen, suggesting that it was located within the plant tissue. Mature plants were irrigated with contaminated water and manure slurry by applying the inoculum directly onto the roots of the plants. Cells of *E. coli* O157:H7 could be found in the edible portions of the lettuce plant for up to 5 days following irrigation. The results of these studies suggest that cells of *E. coli* O157:H7 can become internalized via the root system of growing lettuce plants and be transported to the edible portions of the plant.

A large outbreak of *E. coli* O157:H7 infection linked to the consumption of radish sprouts (Sakai City, Japan, 1996) and numerous outbreaks of salmonellosis linked to alfalfa sprouts (104) have resulted in a great deal of research on the interactions between enteric pathogens and seed sprouts. Internal colonization of radish sprouts by *E. coli* O157:H7 was demonstrated using both fluorescence and scanning electron microscopy (57). *S. enterica* serovar Stanley was shown by confocal scanning laser microscopy to enter intact alfalfa sprouts (36). Neither of these studies investigated the point of entry of these bacteria into the plants, but Itoh et al. (57) presented a scanning electron micrograph showing *E. coli* O157:H7 in the xylems of the radish sprouts. Vascular contamination of bean sprouts by *E. coli* and *S. enterica* was confirmed using surface sterilization and in situ staining (116). The authors of the study suggested that bacteria were able to enter the plants via cracks in the epidermis or fissures resulting from the emergence of lateral roots. These findings were supported by results of confocal microscopy studies on the interactions between *E. coli* K-12, *K. pneumoniae,* various serovars of *S. enterica,* and seedlings of alfalfa (29). Areas of emergence of lateral roots were colonized by all three bacterial species tested on alfalfa seedlings grown in test tubes containing nutrient medium. *K. pneumoniae* and *E. coli* K-12 were the best and worst endophytic colonizers, respectively. Significant differences in the endophytic colonization patterns of various *S. enterica* serovars were observed, indicating strain specificity for entry into plants in this experimental system. In addition, Cooley et al. (23) demonstrated that invasion at the lateral root junction of the experimental model plant *Arabidopsis thaliana* was severely attenuated for *S. enterica* mutants deficient in flagellin production or motility and inoculated onto the seeds. This finding suggested that motility plays an important role in the internalization of this pathogen into plants. Important also is the report by Warriner et al. (116) that showed that bioluminescent *E. coli* cells have the ability to colonize internally the roots of spinach seedlings grown from inoculated seeds in soil microcosms, providing evidence again that internalization is not solely an in vitro phenomenon but may occur in plants in the field. The ability of *E. coli* and *S. enterica* to contaminate the internal portions of seed sprouts and seedlings, thereby escaping inactivation by surface sanitizers, may pose a significant problem in the food industry.

SURVIVAL AND PERSISTENCE

Numerous avenues exist for the introduction of enteric pathogens onto crops in the field. Frequently, fields are fertilized with manure applied prior to planting. Crops are irrigated with water from a variety of sources, such as ponds, streams, and lakes, as well as subsurface well water. Animal husbandry occurs in close proximity to planted fields where contamination through runoff or direct contamination of water or crops may take place. Feral animals may also have easy access to the growing fields. The continuum between planting and harvest is an important area of study in order to understand how certain organisms become associated with preharvest plants. Preharvest contamination by enteric pathogens has been suggested as sufficient to result in the sale of tainted produce. Ample experimental evidence for the ability of enteric bacteria to attach to fruits and vegetables has been provided, as described in the above sections. More recently, research has begun to document the ability of foodborne pathogens to persist once attached to plant surfaces. We present below some of the most relevant findings regarding the persistence of human pathogens on plants, supporting the hypothesis that preharvest bacterial attachment and subsequent survival may result in the consumption of contaminated produce.

E. coli

Recently, there have been attempts to quantify the survival of *E. coli* O157:H7 on growing crops under both laboratory and greenhouse conditions (52, 95, 96). The pathogen was found to survive on inoculated cantaloupe surfaces for longer than 2 weeks in a greenhouse setting (98). The authors of the study found that growth under more humid conditions resulted in enhanced survival of the pathogen.

Wachtel et al. (110) investigated the association of *E. coli* O157:H7 with preharvest lettuce seedlings. Seedlings grown in soil irrigated with contaminated water harbored the pathogen at high levels, even after several days of growth. Large populations of *E. coli* O157:H7 were detected on plant roots, hypocotyls, and cotyledons, regardless of the initial bacterial concentration in the irrigation water. In a similar study, lettuce plants grown to 20, 30, or 40 days of age were spray irrigated with water containing *E. coli* O157:H7. Plants were sampled when they reached maturity. Nine of the 11 plants irrigated with contaminated water harbored the pathogen, even after 20 additional days of growth (95). In a follow-up study, lettuce plants in a growth chamber were spray irrigated either once or intermittently with water containing *E. coli* O157:H7 and then sampled over a 30-day period (96). Plants receiving a single irrigation with 10^4 CFU/ml still harbored the pathogen at harvest. Intermittent irrigation on days 7 and 14 resulted in the recovery of high levels of *E. coli* O157:H7 at harvest (5.0 log CFU).

Ibekwe et al. (52) used culture methods as well as real-time PCR to investigate the fate of *E. coli* O157:H7 in the phyllosphere, rhizosphere, and nonrhizosphere soils surrounding lettuce plants exposed to contaminated irrigation water. Results confirmed previous findings that drip irrigation using 10^7 CFU contained in 1 liter of phosphate-buffered saline could result in plants' harboring of the pathogen. *E. coli* O157:H7 survived for 12 days in the phyllosphere of irrigated lettuce, but levels depended greatly on the type of soil used to cultivate the plants. High levels were also detected in samples from the rhizosphere and nonrhizosphere soils surrounding the plants, raising the concern of recontamination during harvest. The fate of *E. coli* O157:H7 in soil as well as on the tissue of planted carrots and onions has also been investigated (53). Carrots planted into soil amended with manure compost harbored the pathogen at high levels 84 days after transplanting. Survival was somewhat lower on onion tissue, with *E. coli* O157:H7 detectable only by enrichment after 49 days of growth. When the experiment was terminated, high levels of viable cells were detected in the soil surrounding onions and carrots, again underscoring the potential for recontamination of plant tissue during harvest.

The results of the laboratory- and greenhouse-level studies discussed above were confirmed by a field-scale investigation conducted by Islam et al. (55). *E. coli* O157:H7 was found to persist on lettuce (>77 days) and parsley (>177 days) after planting in fields treated with contaminated manure composts or irrigation water. Soils surrounding the plants remained positive for the presence of the pathogen for far longer periods than the plants themselves. Clearly, *E. coli* O157:H7 can survive for extended periods following attachment to plant tissue, regardless of whether cells originate in irrigation water or contaminated manure.

S. enterica

There have been several reports indicating that *S. enterica* is also able to persist following introduction onto growing crops. *S. enterica* serovar Thompson was found to be capable of colonizing cilantro leaves, with bacterial levels increasing throughout 6 days of incubation (11). Incubation of the plants at 30°C increased the levels of *S. enterica*, indicating that warm temperatures may increase the competitiveness of the organism in the cilantro phyllosphere. The population size of the pathogen on the leaves was also greatly dependent on relative humidity.

Islam et al. (54) investigated the survival of *S. enterica* serovar Typhimurium on lettuce and parsley grown in fields treated with contaminated composts or irrigation water. Results indicated that the organisms persisted for 63 days on lettuce and 231 days on parsley. Similar results were demonstrated in field studies on carrots and radishes (56). Additionally, contamination of arugula leaves by serovar Typhimurium was detected following 17 weeks of growth in manure-fertilized soils (80).

The above-described research demonstrates conclusively the ability of enteric pathogens to persist for extended periods of time once attached to plant tissue. This further emphasizes the need to understand in greater detail the mechanisms by which foodborne pathogens attach to plant tissues and the ecology of these bacteria in the plant-soil environment.

CONCLUSION

The survival of pathogenic bacteria in soil, water, and animal wastes has been well established. The recent surge in outbreaks linked to the consumption of fresh produce has led to an intensive research effort to understand the ecology of human pathogens in the plant environment. Even more recently, the persistence of pathogens on and in growing plants has been scrutinized. Contrary to prevailing theory, human pathogens were found to survive on growing plants over prolonged periods of time under field conditions. In addition, the ability of these organisms to infiltrate damaged plant tissues has been recognized.

Much of the available literature on the attachment of bacteria to plants concerns plant pathogens and symbionts. In contrast, the ability of foodborne pathogens to attach to plant surfaces has only begun to be characterized chemically and molecularly. With the increasing realization that human pathogens and plant pathogens have in common many factors and pathways that are essential for their interaction with their hosts, one can hypothesize that human pathogens may utilize, among other strategies, mechanisms of attachment to plant tissue that are similar to those described for plant pathogenic or plant-associated bacteria. Since the majority of postharvest treatments aimed at eliminating or removing pathogens from produce have failed, a better understanding of these mechanisms will be required to devise more effective intervention strategies.

ACKNOWLEDGMENTS

We thank William Fett and Peter Irwin for their valuable comments while reviewing this chapter, Nereus Gunther for preparation of the artwork included in Fig. 1, and Brendan Niemira and Paul Pierlott for assistance with the scanning electron micrograph.

REFERENCES

1. **Ackers, M.-L., B. E. Mahon, E. Leahy, B. Goode, T. Damrow, P. S. Hayes, W. F. Bibb, D. H. Rice, T. J. Barrett, L. Hutwanger, P. M. Griffin, and L. Slutsker.** 1998. An outbreak of *Escherichia coli* O157:H7 infections associated with leaf lettuce consumption. *J. Infect. Dis.* **177**:1588–1593.
2. **Annous, B. A., G. M. Sapers, A. M. Mattrazzo, and D. C. Riordan.** 2001. Efficacy of washing with a commercial flatbed brush washer, using conventional and experimental washing agents, in reducing populations of *Escherichia coli* on artificially inoculated apples. *J. Food Prot.* **64**:159–163.
3. **Anonymous.** 28 January 2003, posting date. *FDA Survey of Domestic Fresh Produce.* FY 2000/2001 field assignment. [Online.] U.S. Food and Drug Administration, Washington, D.C. http://www.cfsan.fda.gov/~dms/prodsu10.html.
4. **Anonymous.** 30 January 2001. *FDA Survey of Imported Fresh Produce.* FY 1999 field assignment. [Online.] U.S. Food and Drug Administration, Washington, D.C. http://www.cfsan.fda.gov/~dms/prodsur6.html.

5. **Ausmees, N., K. Jacobsson, and M. Lindberg.** 2001. A unipolarly located, cell-surface-associated agglutinin, RapA, belongs to a family of *Rhizobium*-adhering proteins (Rap) in *Rhizobium leguminosarum* bv. *trifolii*. *Microbiology* **147:**549–559.
6. **Barak, J. D., L. C. Whitehand, and A. O. Charkowski.** 2002. Differences in attachment of *Salmonella enterica* serovars and *Escherichia coli* O157:H7 to alfalfa sprouts. *Appl. Environ. Microbiol.* **68:**4758–4763.
7. **Bartz, J. A., and R. K. Showalter.** 1981. Infiltration of tomatoes by bacteria in aqueous suspension. *Phytopathology* **71:**515–518.
8. **Beuchat, L. R.** 2002. Ecological factors influencing survival and growth of human pathogens on raw fruits and vegetables. *Microbes Infect.* **4:**413–423.
9. **Bohlool, B. B., and E. L. Schmidt.** 1974. Lectins: a possible basis for specificity in the *Rhizobium*-legume root nodule symbiosis. *Science* **185:**269–271.
10. **Bolton, H., J. K. Fredrickson, and L. F. Elliot.** 1993. Microbial ecology of the rhizosphere, p. 646. *In* F. B. Metting (ed.), *Soil Microbial Ecology.* Marcel Dekker, New York, N.Y.
11. **Brandl, M. T., and R. E. Mandrell.** 2002. Fitness of *Salmonella enterica* serovar Thompson in the cilantro phyllosphere. *Appl. Environ. Microbiol.* **68:**3614–3621.
12. **Brandl, M. T., and J.-M. Monier.** 2005. Methods in microscopy for the visualization of bacteria and their behavior on plants. *In* G. M. Sapers, J. R. Gorny, and A. E. Yousef (ed.), *Microbiology of Fruits and Vegetables.* CRC Press LLC, Ames, Iowa.
13. **Breuer, T., D. H. Benkel, R. L. Shapiro, W. N. Hall, M. M. Winnett, M. J. Linn, J. Neimann, T. J. Barrett, S. Dietrich, F. P. Downes, D. M. Toney, J. L. Pearson, H. Rolka, L. Slutsker, and P. M. Griffin.** 2001. A multistate outbreak of *Escherichia coli* O157:H7 infections linked to alfalfa sprouts grown from contaminated seeds. *Emerg. Infect. Dis.* **7:**977–982.
14. **Buchanan, R. L., S. G. Edelson, R. L. Miller, and G. M. Sapers.** 1999. Contamination of intact apples after immersion in an aqueous environment containing *Escherichia coli* O157:H7. *J. Food Prot.* **62:**444–450.
15. **Buck, J. W., R. R. Walcott, and L. R. Beuchat.** 21 January 2003, posting date. Recent trends in microbiological safety of fruits and vegetables. *Plant Health Prog.* [Online.] doi:10.1094/PHP-2003-0121-01-RV.
16. **Burdman, S., G. Dulguerova, Y. Okon, and E. Jurkevitch.** 2001. Purification of the major outer membrane protein of *Azospirillum brasilense*, its affinity to plant roots, and its involvement in cell aggregation. *Mol. Plant-Microbe Interact.* **14:**555–561.
17. **Burnett, S. L., and L. R. Beuchat.** 2000. Human pathogens associated with raw produce and unpasteurized juices, and difficulties in decontamination. *J. Ind. Microbiol. Biotechnol.* **25:**281–287.
18. **Burnett, S. L., J. Chen, and L. R. Beuchat.** 2000. Attachment of *Escherichia coli* O157:H7 to the surfaces and internal structures of apples as detected by confocal scanning laser microscopy. *Appl. Environ. Microbiol.* **66:**4679–4687.
19. **Campbell, V. J., J. Mohle-Boetani, R. Reporter, S. Abbott, J. Farrar, M. T. Brandl, R. E. Mandrell, and S. B. Werner.** 2001. An outbreak of *Salmonella* serotype Thompson associated with fresh cilantro. *J. Infect. Dis.* **183:**984–987.
20. **Centers for Disease Control and Prevention.** 2003. *Foodborne Outbreaks Due to Bacterial Etiologies,* 2003. [Online.] Centers for Disease Control and Prevention, Atlanta, Ga. http://www.cdc.gov/foodborneoutbreaks/us_outb/fbo2003/bacterial03.pdf.

21. **Charkowski, A. O., J. D. Barak, C. Z. Sarreal, and R. E. Mandrell.** 2002. Differences in growth of *Salmonella enterica* and *Escherichia coli* O157:H7 on alfalfa sprouts. *Appl. Environ. Microbiol.* **68:**3114–3120.
22. **Clark, F. E.** 1949. Soil microorganisms and plant roots. *Adv. Agron.* **1:**241–288.
23. **Cooley, M. B., W. G. Miller, and R. E. Mandrell.** 2003. Colonization of *Arabidopsis thaliana* with *Salmonella enterica* or enterohemorrhagic *Escherichia coli* O157:H7 and competition by *Enterobacter asburiae. Appl. Environ. Microbiol.* **69:**4915–4926.
24. **Croes, C. L., S. Moens, E. Van Bastelaere, J. Vanderleyden, and K. W. Michiels.** 1993. The polar flagellum mediates *Azospirillum brasilense* adsorption to wheat roots. *J. Gen. Microbiol.* **139:**2261–2269.
25. **Curl, E. A., and B. Truelove.** 1986. Root exudates, p. 55–92. *In* B. Yaron (ed.), *The Rhizosphere.* Springer-Verlag, Berlin, Germany.
26. **Curl, E. A., and B. Truelove.** 1986. The structure and function of roots, p. 9–54. *In* B. Yaron (ed.), *The Rhizosphere.* Springer-Verlag, Berlin, Germany.
27. **Del Gallo, M., M. Negi, and C. A. Neyra.** 1989. Calcofluor- and lectin-binding exocellular polysaccharides of *Azospirillum brasilense* and *Azospirillum lipoferum. J. Bacteriol.* **171:**3504–3510.
28. **Derridj, S.** 1996. Nutrients on the leaf surface, p. 25–42. *In* C. E. Morris, P. C. Nicot, and C. Nguyen-The (ed.), *Aerial Plant Surface Microbiology.* Plenum Press, New York, N.Y.
29. **Dong, Y., A. L. Iniguez, B. M. Ahmer, and E. W. Triplett.** 2003. Kinetics and strain specificity of rhizosphere and endophytic colonization by enteric bacteria on seedlings of *Medicago sativa* and *Medicago truncatula. Appl. Environ. Microbiol.* **69:**1783–1790.
30. **Eblen, B. S., M. O. Walderhaug, S. Edelson-Mammel, S. J. Chirtel, A. De Jesus, R. I. Merker, R. L. Buchanan, and A. J. Miller.** 2004. Potential for internalization, growth, and survival of *Salmonella* and *Escherichia coli* O157:H7 in oranges. *J. Food Prot.* **67:**1578–1584.
31. **Esmelindro, A. A., S. Girardi Jdos, A. Mossi, R. A. Jacques, and C. Dariva.** 2004. Influence of agronomic variables on the composition of mate tea leaves (*Ilex paraguariensis*) extracts obtained from CO_2 extraction at 30 degrees C and 175 bar. *J. Agric. Food Chem.* **52:**1990–1995.
32. **Fenlon, D. R., I. D. Ogden, A. Vinten, and I. Svoboda.** 2000. The fate of *Escherichia coli* and *E. coli* O157 in cattle slurry after application to land. *Symp. Ser. Soc. Appl. Microbiol.* **88:**149S–156S.
33. **Fukushima, H., K. Hoshina, and M. Gomyoda.** 1999. Long-term survival of Shiga toxin-producing *Escherichia coli* O26, O111, and O157 in bovine feces. *Appl. Environ. Microbiol.* **65:**5177–5181.
34. **Gagliardi, J. V., and J. S. Karns.** 2002. Persistence of *Escherichia coli* O157:H7 in soil and on plant roots. *Environ. Microbiol.* **4:**89–96.
35. **Gal, M., G. M. Preston, R. C. Massey, A. J. Spiers, and P. B. Rainey.** 2003. Genes encoding a cellulosic polymer contribute toward the ecological success of *Pseudomonas fluorescens* SBW25 on plant surfaces. *Mol. Ecol.* **12:**3109–3121.
36. **Gandhi, M., S. Golding, S. Yaron, and K. R. Matthews.** 2001. Use of green fluorescent protein expressing *Salmonella* Stanley to investigate survival, spatial location, and control on alfalfa sprouts. *J. Food Prot.* **64:**1891–1898.
37. **Gelvin, S. B.** 2003. *Agrobacterium*-mediated plant transformation: the biology behind the "gene-jockeying" tool. *Microbiol. Mol. Biol. Rev.* **67:**16–37.

38. **Gorski, L., J. D. Palumbo, and R. E. Mandrell.** 2003. Attachment of *Listeria monocytogenes* to radish tissue is dependent upon temperature and flagellar motility. *Appl. Environ. Microbiol.* **69**:258–266.
39. **Gorski, L., J. D. Palumbo, and K. D. Nguyen.** 2004. Strain-specific differences in the attachment of *Listeria monocytogenes* to alfalfa sprouts. *J. Food Prot.* **67**:2488–2495.
40. **Guo, X., J. Chen, R. E. Brackett, and L. R. Beuchat.** 2001. Survival of salmonellae on and in tomato plants from the time of inoculation at flowering and early stages of fruit development through fruit ripening. *Appl. Environ. Microbiol.* **67**:4760–4764.
41. **Guo, X., M. W. van Iersel, J. Chen, R. E. Brackett, and L. R. Beuchat.** 2002. Evidence of association of salmonellae with tomato plants grown hydroponically in inoculated nutrient solution. *Appl. Environ. Microbiol.* **68**:3639–3643.
42. **Haahtela, K., E. Tarkka, and T. K. Korhonen.** 1985. Type 1 fimbria-mediated adhesion of enteric bacteria to grass roots. *Appl. Environ. Microbiol.* **49**:1182–1185.
43. **Han, Y., D. M. Sherman, R. H. Linton, S. S. Nielsen, and P. E. Nielsen.** 2000. The effects of washing and chlorine dioxide gas on survival and attachment of *Escherichia coli* O157:H7 to green pepper surfaces. *Food Microbiol.* **17**:521–533.
44. **Handelsman, J., and E. V. Stabb.** 1996. Biocontrol of soilborne plant pathogens. *Plant Cell* **8**:1855–1869.
45. **Hassan, A. N., and J. F. Frank.** 2004. Attachment of *Escherichia coli* O157:H7 grown in tryptic soy broth and nutrient broth to apple and lettuce surfaces as related to cell hydrophobicity, surface charge, and capsule production. *Int. J. Food Microbiol.* **96**:103–109.
46. **Hassan, A. N., and J. F. Frank.** 2003. Influence of surfactant hydrophobicity on the detachment of *Escherichia coli* O157:H7 from lettuce. *Int. J. Food Microbiol.* **87**:145–152.
47. **Hilborn, E. D., J. H. Mermin, P. A. Mshar, J. L. Hadler, A. Voetsch, C. Wojtkunski, M. Swartz, R. Mshar, M. A. Lambert-Fair, J. A. Farrar, M. K. Glynn, and L. Slutsker.** 1999. A multistate outbreak of *Escherichia coli* O157:H7 infections associated with consumption of mesclun lettuce. *Arch. Intern. Med.* **159**:1758–1764.
48. **Hirano, S. S., L. S. Baker, and C. D. Upper.** 1996. Raindrop momentum triggers growth of leaf-associated populations of *Pseudomonas syringae* on field-grown snap bean plants. *Appl. Environ. Microbiol.* **62**:2560–2566.
49. **Hirano, S. S., and C. D. Upper.** 2000. Bacteria in the leaf ecosystem with emphasis on *Pseudomonas syringae*—a pathogen, ice nucleus, and epiphyte. *Microbiol. Mol. Biol. Rev.* **64**:624–653.
50. **Ho, S. C., J. L. Wang, and M. Schindler.** 1990. Carbohydrate binding activities of *Bradyrhizobium japonicum*. I. Saccharide-specific inhibition of homotypic and heterotypic adhesion. *J. Cell Biol.* **111**:1631–1638.
51. **Huang, J.-S.** 1986. Ultrastructure of bacterial penetration in plants. *Annu. Rev. Phytopathol.* **24**:141–157.
52. **Ibekwe, A. M., P. M. Watt, P. J. Shouse, and C. M. Grieve.** 2004. Fate of *Escherichia coli* O157:H7 in irrigation water on soils and plants as validated by culture method and real-time PCR. *Can. J. Microbiol.* **50**:1007–1014.
53. **Islam, M., J. Morgan, M. P. Doyle, and X. Jiang.** 2004. Fate of *Escherichia coli* O157:H7 in manure compost-amended soil and on carrots and onions grown in an environmentally controlled growth chamber. *J. Food Prot.* **67**:574–578.

54. **Islam, M., J. Morgan, M. P. Doyle, S. C. Phatak, P. Millner, and X. Jiang.** 2004. Persistence of *Salmonella enterica* serovar Typhimurium on lettuce and parsley and in soils on which they were grown in fields treated with contaminated manure composts or irrigation water. *Foodborne Pathog. Dis.* **1**:27–35.
55. **Islam, M., M. P. Doyle, S. C. Phatak, M. Millner, and X. Jiang.** 2004. Persistence of enterohemorrhagic *Escherichia coli* O157:H7 in soil and on leaf lettuce and parsley grown in fields treated with contaminated manure composts or irrigation water. *J. Food Prot.* **67**:1365–1370.
56. **Islam, M., J. Morgan, M. P. Doyle, S. C. Phatak, P. Millner, and X. Jiang.** 2004. Fate of *Salmonella enterica* serovar Typhimurium on carrots and radishes grown in fields treated with contaminated manure composts or irrigation water. *Appl. Environ. Microbiol.* **70**:2497–2502.
57. **Itoh, Y., Y. Sugita-Konishi, F. Kasuga, M. Iwaki, Y. Hara-Kudo, N. Saito, Y. Noguchi, H. Konuma, and S. Kumagai.** 1998. Enterohemorrhagic *Escherichia coli* O157:H7 present in radish sprouts. *Appl. Environ. Microbiol.* **64**:1532–1535.
58. **Jaeger, C. H., S. E. Lindow, S. Miller, E. Clark, and M. K. Firestone.** 1999. Mapping of sugar and amino acid availability in soil around roots with bacterial sensors of sucrose and tryptophan. *Appl. Environ. Microbiol.* **65**:2685–2690.
59. **Jeffree, C. E.** 1996. Structure and ontogeny of plant cuticles, p. 33–82. *In* G. Kerstiens (ed.), *Plant Cuticles.* Bios, Oxford, United Kingdom.
60. **Jiang, X., J. Morgan, and M. P. Doyle.** 2002. Fate of *Escherichia coli* O157:H7 in manure-amended soil. *Appl. Environ. Microbiol.* **68**:2605–2609.
61. **Joyner, D. C., and S. E. Lindow.** 2000. Heterogeneity of iron bioavailability on plants assessed with a whole-cell GFP-based bacterial biosensor. *Microbiology* **146**:2435–2445.
62. **Kenney, S. J., S. L. Burnett, and L. R. Beuchat.** 2001. Location of *Escherichia coli* O157:H7 on and in apples as affected by bruising, washing, and rubbing. *J. Food Prot.* **64**:1328–1333.
63. **Korhonen, T. K., E. Tarkka, H. Ranta, and K. Haahtela.** 1983. Type 3 fimbriae of *Klebsiella* sp.: molecular characterization and role in bacterial adhesion to plant roots. *J. Bacteriol.* **155**:860–865.
64. **Leben, C.** 1988. Relative humidity and the survival of epiphytic bacteria with buds and leaves of cucumber plants. *Phytopathology* **78**:179–185.
65. **Leveau, J. H. J., and S. E. Lindow.** 2001. Appetite of an epiphyte: quantitative monitoring of bacterial sugar consumption in the phyllosphere. *Proc. Natl. Acad. Sci. USA* **98**:3446–3453.
66. **Li, Y., R. E. Brackett, J. Chen, and L. R. Beuchat.** 2001. Survival and growth of *Escherichia coli* O157:H7 inoculated onto cut lettuce before or after heating in chlorinated water, followed by storage at 5 or 15 degrees C. *J. Food Prot.* **64**:305–309.
67. **Liao, C. H., and G. M. Sapers.** 2000. Attachment and growth of *Salmonella* Chester on apple fruits and in vivo response of attached bacteria to sanitizer treatments. *J. Food Prot.* **63**:876–883.
68. **Lindow, S. E., and M. T. Brandl.** 2003. Microbiology of the phyllosphere. *Appl. Environ. Microbiol.* **69**:1875–1883.
69. **Loh, J. T., S. C. Ho, A. W. de Feijter, J. L. Wang, and M. Schindler.** 1993. Carbohydrate binding activities of *Bradyrhizobium japonicum:* unipolar localization of the lectin BJ38 on the bacterial cell surface. *Proc. Natl. Acad. Sci. USA* **90**:3033–3037.

70. **Mahon, B. E., A. Ponka, W. N. Hall, K. Komatsu, S. E. Dietrich, A. Siitonen, G. Cage, P. S. Hayes, M. A. Lambert-Fair, N. H. Bean, P. M. Griffin, and L. Slutsker.** 1997. An international outbreak of *Salmonella* infections caused by alfalfa sprouts grown from contaminated seeds. *J. Infect. Dis.* **175**:876–882.
71. **Mandrell, R. E., M. A. Apicella, R. Lindstedt, and H. Leffler.** 1994. Possible interaction between animal lectins and bacterial carbohydrates. *Methods Enzymol.* **236**:231–254.
72. **Mandrell, R. E., L. Gorski, and M. Brandl.** 2005. Attachment of microorganisms to fresh produce. *In* G. M. Sapers, J. R. Gorny, and A. E. Yousef (ed.), *Microbiology of Fruits and Vegetables*. CRC Press, Ames, Iowa.
73. **Matthysse, A. G.** 1987. Characterization of nonattaching mutants of *Agrobacterium tumefaciens*. *J. Bacteriol.* **169**:313–323.
74. **Matthysse, A. G., K. V. Holmes, and R. H. G. Gulrlitz.** 1981. Elaboration of cellulose fibrils by *Agrobacterium tumefaciens* during attachment on carrot cells. *J. Bacteriol.* **145**:583–595.
75. **Matthysse, A. G., and S. McMahan.** 2001. The effect of the *Agrobacterium tumefaciens attR* mutation on attachment and root colonization differs between legumes and other dicots. *Appl. Environ. Microbiol.* **67**:1070–1075.
76. **Mechaber, W. L., D. B. Marshall, R. A. Mechaber, R. T. Jobe, and F. S. Chew.** 1996. Mapping leaf surface landscapes. *Proc. Natl. Acad. Sci. USA* **93**:4600–4603.
77. **Mercier, J., and S. E. Lindow.** 2000. Role of leaf surface sugars in colonization of plants by bacterial epiphytes. *Appl. Environ. Microbiol.* **66**:369–374.
78. **Miller, W. G., M. T. Brandl, B. Quinones, and S. E. Lindow.** 2001. Biological sensor for sucrose availability: relative sensitivities of various reporter genes. *Appl. Environ. Microbiol.* **67**:1308–1317.
79. **Mohle-Boetani, J. C., J. A. Farrar, S. B. Werner, D. Minassian, R. Bryant, S. Abbott, L. Slutsker, and D. J. Vugia.** 2001. *Escherichia coli* O157 and *Salmonella* infections associated with sprouts in California, 1996–1998. *Ann. Intern. Med.* **135**:239–247.
80. **Natvig, E. E., S. C. Ingham, B. H. Ingham, L. R. Cooperband, and T. R. Roper.** 2002. *Salmonella enterica* serovar Typhimurium and *Escherichia coli* contamination of root and leaf vegetables grown in soils with incorporated bovine manure. *Appl. Environ. Microbiol.* **68**:2737–2744.
81. **Ojanen-Reuhs, T., N. Kalkkinen, B. Westerlund-Wikstrom, J. van Doorn, K. Haahtela, E. L. Nurmiaho-Lassila, K. Wengelnik, U. Bonas, and T. K. Korhonen.** 1997. Characterization of the *fimA* gene encoding bundle-forming fimbriae of the plant pathogen *Xanthomonas campestris* pv. vesicatoria. *J. Bacteriol.* **179**:1280–1290.
82. **Penteado, A. L., B. S. Eblen, and A. J. Miller.** 2004. Evidence of *Salmonella* internalization into fresh mangos during simulated postharvest insect disinfestation procedures. *J. Food Prot.* **67**:181–184.
83. **Quadt-Hallman, A., N. Benhamou, and J. W. Kloepper.** 1997. Bacterial endophytes in cotton: mechanisms of entering the plant. *Can. J. Microbiol.* **43**:577–582.
84. **Reina, L. D., H. P. Fleming, and F. Breidt, Jr.** 2002. Bacterial contamination of cucumber fruit through adhesion. *J. Food Prot.* **65**:1881–1887.
85. **Reuhs, B. L., J. S. Kim, and A. G. Matthysse.** 1997. Attachment of *Agrobacterium tumefaciens* to carrot cells and *Arabidopsis* wound sites is correlated with the presence of a cell-associated, acidic polysaccharide. *J. Bacteriol.* **179**:5372–5379.

86. **Rodgers, S. L., J. N. Cash, M. Siddiq, and E. T. Ryser.** 2004. A comparison of different chemical sanitizers for inactivating *Escherichia coli* O157:H7 and *Listeria monocytogenes* in solution and on apples, lettuce, strawberries, and cantaloupe. *J. Food Prot.* **67**:721–731.

87. **Rojas, C. M., J. H. Ham, W. L. Deng, J. J. Doyle, and A. Collmer.** 2002. HecA, a member of a class of adhesins produced by diverse pathogenic bacteria, contributes to the attachment, aggregation, epidermal cell killing, and virulence phenotypes of *Erwinia chrysanthemi* EC16 on *Nicotiana clevelandii* seedlings. *Proc. Natl. Acad. Sci. USA* **99**:13142–13147.

88. **Romantschuk, M.** 1992. Attachment of plant pathogenic bacteria to plant surfaces. *Annu. Rev. Phytopathol.* **30**:225–243.

89. **Romantschuk, M.** 2004. Bacterial attachment to leaves, p. 75–78. *In* R. M. Goodman (ed.), *Encyclopedia of Plant and Crop Science*. Marcel Dekker, Inc., New York, N.Y.

90. **Romantschuk, M., and D. H. Bamford.** 1986. The causal agent of halo blight in bean, *Pseudomonas syringae* pv. phaseolicola, attaches to stomata via its pili. *Microb. Pathog.* **1**:139–148.

91. **Sanders, L. C., C. S. Wang, L. L. Walling, and E. M. Lord.** 1991. A homolog of the substrate adhesion molecule vitronectin occurs in four species of flowering plants. *Plant Cell* **3**:629–635.

92. **Seo, K. H., and J. F. Frank.** 1999. Attachment of *Escherichia coli* O157:H7 to lettuce leaf surface and bacterial viability in response to chlorine treatment as demonstrated by using confocal scanning laser microscopy. *J. Food Prot.* **62**:3–9.

93. **Smit, G., J. W. Kijne, and B. J. Lugtenberg.** 1987. Involvement of both cellulose fibrils and a Ca^{2+}-dependent adhesin in the attachment of *Rhizobium leguminosarum* to pea root hair tips. *J. Bacteriol.* **169**:4294–4301.

94. **Smit, G., J. W. Kijne, and B. J. Lugtenberg.** 1989. Roles of flagella, lipopolysaccharide, and a Ca^{2+}-dependent cell surface protein in attachment of *Rhizobium leguminosarum* biovar viciae to pea root hair tips. *J. Bacteriol.* **171**:569–572.

95. **Solomon, E. B., C. J. Potenski, and K. R. Matthews.** 2002. Effect of irrigation method on transmission to and persistence of *Escherichia coli* O157:H7 on lettuce. *J. Food Prot.* **65**:673–676.

96. **Solomon, E. B., H.-J. Pang, and K. R. Matthews.** 2003. Persistence of *Escherichia coli* O157:H7 on lettuce plants following spray irrigation with contaminated water. *J. Food Prot.* **66**:2198–2202.

97. **Solomon, E. B., S. Yaron, and K. R. Matthews.** 2002. Transmission of *Escherichia coli* O157:H7 from contaminated manure and irrigation water to lettuce plant tissue and its subsequent internalization. *Appl. Environ. Microbiol.* **68**:397–400.

98. **Stine, S. W., I. Song, C. Y. Choi, and C. P. Gerba.** 2003. Effect of environmental conditions on the survival of microbial pathogens on the surface of cantaloupe, abstr. P-092, p. 125. *Abstr. 103rd Gen. Meet. Am. Soc. Microbiol.* American Society for Microbiology, Washington, D.C.

99. **Sudakevitz, D., A. Imberty, and N. Gilboa-Garber.** 2002. Production, properties and specificity of a new bacterial L-fucose- and D-arabinose-binding lectin of the plant aggressive pathogen *Ralstonia solanacearum*, and its comparison to related plant and microbial lectins. *J. Biochem.* **132**:353–358.

100. **Sudakevitz, D., N. Kostlanova, G. Blatman-Jan, E. P. Mitchell, B. Lerrer, M. Wimmerova, D. J. Katcoff, A. Imberty, and N. Gilboa-Garber.** 2004. A new *Ralstonia solanacearum* high-affinity mannose-binding lectin RS-IIL structurally resembling the *Pseudomonas aeruginosa* fucose-specific lectin PA-IIL. *Mol. Microbiol.* **52**:691–700.

101. **Swart, S., B. J. Lugtenberg, G. Smit, and J. W. Kijne.** 1994. Rhicadhesin-mediated attachment and virulence of an *Agrobacterium tumefaciens chvB* mutant can be restored by growth in a highly osmotic medium. *J. Bacteriol.* **176**:3816–3819.

102. **Takeuchi, K., and J. F. Frank.** 2000. Penetration of *Escherichia coli* O157:H7 into lettuce tissues as affected by inoculum size and temperature and the effect of chlorine treatment on cell viability. *J. Food Prot.* **63**:434–440.

103. **Takeuchi, K., C. M. Matute, A. N. Hassan, and J. F. Frank.** 2000. Comparison of the attachment of *Escherichia coli* O157:H7, *Listeria monocytogenes*, *Salmonella typhimurium*, and *Pseudomonas fluorescens* to lettuce leaves. *J. Food Prot.* **63**:1433–1437.

104. **Taormina, P. J., L. R. Beuchat, and L. Slutsker.** 1999. Infections associated with eating seed sprouts: an international concern. *Emerg. Infect. Dis.* **5**:626–634.

105. **Tsien, H. C., and E. L. Schmidt.** 1981. Localization and partial characterization of soybean lectin-binding polysaccharide of *Rhizobium japonicum*. *J. Bacteriol.* **145**:1063–1074.

106. **Ukuku, D. O., and W. Fett.** 2002. Behavior of *Listeria monocytogenes* inoculated on cantaloupe surfaces and efficacy of washing treatments to reduce transfer from rind to fresh-cut pieces. *J. Food Prot.* **65**:924–930.

107. **Ukuku, D. O., and W. F. Fett.** 2002. Relationship of cell surface charge and hydrophobicity to strength of attachment of bacteria to cantaloupe rind. *J. Food Prot.* **65**:1093–1099.

108. **Vesper, S. J.** 1987. Production of pili (fimbriae) by *Pseudomonas fluorescens* and correlation with attachment to corn roots. *Appl. Environ. Microbiol.* **53**:1397–1405.

109. **Wachtel, M., and A. Charkowski.** 2002. Cross-contamination of lettuce with *Escherichia coli* O157:H7. *J. Food Prot.* **65**:465–470.

110. **Wachtel, M. R., L. C. Whitehand, and R. E. Mandrell.** 2002. Association of *Escherichia coli* O157:H7 with preharvest leaf lettuce upon exposure to contaminated irrigation water. *J. Food Prot.* **65**:18–25.

111. **Wagner, V. T., and A. G. Matthysse.** 1992. Involvement of a vitronectin-like protein in attachment of *Agrobacterium tumefaciens* to carrot suspension culture cells. *J. Bacteriol.* **174**:5999–6003.

112. **Waisel, Y., A. Eshel, and U. Kafkafi.** 1996. *Plant Roots: the Hidden Half*, 2nd ed. Marcel Dekker, New York, N.Y.

113. **Wang, G., T. Zhao, and M. P. Doyle.** 1996. Fate of enterohemorrhagic *Escherichia coli* O157:H7 in bovine feces. *Appl. Environ. Microbiol.* **62**:2567–2570.

114. **Wang, G., and M. P. Doyle.** 1998. Survival of enterohemorrhagic *Escherichia coli* O157:H7 in water. *J. Food Prot.* **61**:662–667.

115. **Warriner, K., F. Ibrahim, M. Dickinson, C. Wright, and W. M. Waites.** 2003. Internalization of human pathogens within growing salad vegetables. *Biotechnol. Genet. Eng. Rev.* **20**:117–134.

116. **Warriner, K., S. Spahiolas, M. Dickinson, C. Wright, and W. M. Waites.** 2003. Internalization of bioluminescent *Escherichia coli* and *Salmonella* Montevideo in growing bean sprouts. *J. Appl. Microbiol.* **95:**719–727.

117. **Wei, C. I., T. S. Huang, J. M. Kim, W. F. Lin, M. L. Tamplin, and J. A. Bartz.** 1995. Growth and survival of *Salmonella montevideo* on tomatoes and disinfection with chlorinated water. *J. Food Prot.* **58:**829–836.

118. **Whatley, M. H., J. S. Bodwin, B. B. Lippincott, and J. A. Lippincott.** 1976. Role of *Agrobacterium* cell envelope lipopolysaccharide in infection site attachment. *Infect. Immun.* **13:**1080–1083.

119. **Wilson, M., and S. E. Lindow.** 1994. Coexistence among epiphytic bacterial populations mediated through nutritional resource partitioning. *Appl. Environ. Microbiol.* **60:**4468–4477.

120. **Zhuang, R.-Y., L. R. Beuchat, and F. J. Angulo.** 1995. Fate of *Salmonella montevideo* on and in raw tomatoes as affected by temperature and treatment with chlorine. *Appl. Environ. Microbiol.* **61:**2127–2131.

121. **Zogaj, X., W. Bokranz, M. Nimtz, and U. Romling.** 2003. Production of cellulose and curli fimbriae by members of the family *Enterobacteriaceae* isolated from the human gastrointestinal tract. *Infect. Immun.* **71:**4151–4158.

122. **Zogaj, X., M. Nimtz, M. Rohde, W. Bokranz, and U. Romling.** 2001. The multicellular morphotypes of *Salmonella Typhimurium* and *Escherichia coli* produce cellulose as the second component of the extracellular matrix. *Mol. Microbiol.* **39:**1452–1463.

Microbiology of Fresh Produce
Edited by Karl R. Matthews

Postharvest Handling and Processing: Sources of Microorganisms and Impact of Sanitizing Procedures

4

Jorge M. Fonseca

The conventional belief that high microbial populations in produce, intact or minimally processed, can be reduced to safe levels with disinfecting technology that is presently available is examined in this chapter. An ideal sanitizer is one used before product packaging that can eliminate or reduce significantly levels of undesirable microorganisms associated with the product. The concept of a highly efficacious disinfection strategy is supported by laboratory studies using various sanitizing agents and methods that have yielded impressive results, some even with reduction to undetectable levels. However, in practice, questions remain. To what extent do decontamination steps reduce the risks of consuming produce contaminated with human pathogens? Can we rely on postharvest disinfecting technology to provide safe produce for consumers? What is preventing scientists from developing feasible technology that can be used as standardized killing steps for the various applications in the produce industry? Food safety is a worldwide concern, and the research reported during the last few years seems to be heading in various directions (199). Part of the dilemma centers around the inconsistency in results that have been reported for postharvest sanitizing strategies.

MICROBIOLOGY OF PRODUCE POSTHARVEST

Knowing the initial microbial population and composition on a fruit or vegetable at harvest is important when determining what sanitizing practices should be implemented. The microbial population is generally diverse even within the same crop. Total counts on products after processing range from

Jorge M. Fonseca, Yuma Agricultural Center, The University of Arizona, Yuma, AZ 85364.

10^3 to 10^6 CFU/g (128). In most cases, the bacteria present on a commodity postharvest are similar to those found in the field. Likewise, microbial decay of minimally processed produce is associated with the growth of microorganisms that originates in preharvest environments (32, 58). The microflora of produce is largely composed of *Erwinia herbicola*, *Enterobacter agglomerans*, and *Flavobacterium, Pseudomonas,* and *Xanthomonas* spp. Lactic acid bacteria, *Leuconostoc mesenteroides* and *Lactobacillus* spp., and several species of yeast are also commonly found, especially in fruits (94, 201).

The need for an efficient decontamination system becomes more important when practices and conditions in the field support high levels of microorganisms that can potentially contaminate a product. Events occurring before the crop is planted can affect bacteriological quality and safety of the fruits and vegetables reaching the consumers. Therefore, knowledge of prior use of the land is an important factor to consider. Fields used for livestock production and those where animals frequently roamed are more likely to be contaminated with enteric pathogens (186). Although it is rare, some pathogenic bacteria can survive in soils or sewage sludge for months (135, 194). Crops in the field can support the growth of certain microorganisms, thereby influencing the soil microflora (86). Another factor that can potentially elevate the microbial population or affect the microbiological quality of croplands is continuous flooding, especially if the land is near a livestock production facility. Floodwater can become polluted with animal waste and carry the contaminants downstream, where they may be spread over cropland (33). *Salmonella enterica* serovar Thompson, an indicator used to model *Salmonella enterica*, has the ability to tolerate desiccation and to recover efficiently from water stress (34). Contamination in the field can become a risk of high magnitude if the microorganism gains access to internal areas of the plant through leaf lesions, since internalized pathogens are difficult to reach with sanitizers (161). Internalization of *Salmonella* in tomato tissues occurred when fruits remained in constant contact with contaminated soil for 10 days (79).

The type of irrigation practice (spray versus flood) and the quality of the water source can directly impact the microbial quality of produce at harvest. Pathogens associated with irrigation water can contaminate the surfaces and internal tissues of crops. Internal contamination of lettuce with *Escherichia coli* has been observed following overhead irrigation in greenhouse conditions (177). Excessive moisture at harvest can lead to poor microbial quality of produce. Increased microbial populations in lettuce grown commercially have been observed in lettuce harvested after a rainfall event or shortly after irrigation termination, especially if irrigation was performed with overhead sprinklers (64).

Temperature and light intensity may affect the survival rate of bacteria on the crop tissue. Bacterial populations on plant tissue decline with time, the result of desiccation and exposure to UV light from the sun. Under dry conditions and intensive light, populations of mesophilic bacteria on outer leaves of head lettuce decline rapidly whereas populations of bacteria on leaves within the head remain high for an extended period (64). Moreover, the solar UV-B radiation affects some bacteria more than others. Populations of pathogenic bacteria, such as *E. coli* O157, declined by only 50% after 4 days under dry field conditions (93), which raises concerns due to the low infective dose of this pathogen.

No clear differences between the microbial qualities of organically grown and conventional produce have been demonstrated. Organic produce is a concern because decontamination by irradiation, antimicrobial agents, chemical washes, and other synthetic disinfectants is prohibited in organic production. Accepted practices, for instance, pasteurization and the use of chlorinated water, are optional. Disease outbreaks have been associated with organic produce. Verotoxin-producing *Citrobacter freundii* was found in organically grown parsley (189). It is unknown whether organic practices were the cause since contamination of produce could have occurred at any point during production and postharvest handling. The extensive use of manure in organic production and possible negligence in practicing proper usage regulations are thought to account for the higher risk of contamination of organically grown produce (119). A recent study showed that the incidence of *E. coli* in organic produce was over 5 times higher than that in conventional food; however, the incidence of *E. coli* in certified organic produce was similar to that in conventional food (126).

The ability of a microorganism to grow is influenced by a range of intrinsic and extrinsic factors. For example, bacteria associated with bagged fresh-cut products behave differently in the high-CO_2-concentration atmosphere often used. Indeed, the conditions can favor the growth of lactic acid bacteria and limit pathogen growth. More studies are needed to elucidate the type of produce and conditions that enhance the growth of lactic acid bacteria that create an environment inhibitory for other bacteria (67). *Pseudomonas* spp. are capable of excreting cell wall-degrading enzymes that result in the release of nutrients from plant cells (112). High concentrations of CO_2 suppress *Pseudomonas* growth and likely diminish nutrient availability for other microorganisms. *Enterobacter* has been shown to compete with *Listeria monocytogenes* in lettuce (66). Biostimulants or growth regulators, elicitors and inhibitors, may affect microbial growth on plants. Proteins extracted from bacteria induce plant resistance to a broad array of plant pathogens, including *Pseudomonas syringae* (51). Perhaps stimulation of the plant defense system also provides protection against foodborne pathogens. Research demonstrates that produce with high numbers of indigenous bacteria have very low counts, or undetectable levels, of pathogenic bacteria and thermophilic coliforms (96, 188).

HANDLING PRACTICES PRIOR TO PACKAGING AND STORAGE

Foodborne microbes are often transmitted from workers to produce (124). Lack of worker personal hygiene is perhaps one of the most important factors resulting in contaminated produce's reaching retail markets. Investigation of an outbreak of cholera associated with sliced melon revealed that workers were the likely source of the contamination (1). Many agricultural workers, in part influenced by cultural backgrounds (8), assume that since raw produce originates in the soil it makes no difference whether the harvesting equipment is properly cleaned and maintained (33). Moreover, consumers often undervalue the importance of safe handling, committing mistakes repeatedly and thereby increasing the risk of foodborne illnesses (7, 114). Recommended handling temperatures are rarely maintained throughout the entire postharvest continuum, from immediately postharvest to consumption. *Salmonella* grew more rapidly on cut surfaces of melons that were held at room temperature than on those held at refrigeration temperature (72). The survival rate of *E. coli* 0157:H7 in apples stored below 10°C was lower than that at room temperature (63). The consumer must play an active role in ensuring the safety of produce by storing products properly.

Proper postharvest handling becomes more critical with produce that has irregular surfaces or wounds. Injuries to produce through physical abuse at any stage of production and handling may permit the entry of pathogens (e.g., *E. coli* O157:H7 [149]). Injured tissue can harbor bacteria, limiting access of the sanitizing agent. In a study with Lollo Rosso lettuce, washing with chlorine resulted in a reduction in the microbial population of only 2 log CFU/g (4). Limited reduction in bacterial populations on produce following treatment with a sanitizing agent has been reported (2, 71, 101). The numbers of psychrotrophic bacteria on a product start to increase when the chilling process starts at the warehouse or processing facility, regardless of whether the product has been treated with a sanitizer. The different steps in the operation, including shredding, washing, rinsing, centrifugation, and storage, are processes during which bacteria can proliferate (4).

Microbial Quality and Shelf Life

Besides the potential reduction of microbial populations on produce, sanitizing systems may also affect quality attributes, such as color and visual appearance. Consumers assess appearance and aroma to determine produce quality. This practice, although effective in some cases, may be a risky approach as pathogenic bacteria are not always associated with tissue deterioration. Microbial populations may exceed safe levels on produce of acceptable visual quality since technology that extends shelf life may provide more time for pathogenic bacteria to grow (33). Fermentation and decay, demonstrated by off odor and soft rot, respectively, do not necessarily correlate with microbial counts. High visual quality ratings were given to fresh-cut lettuce even though populations of psychrotrophic microorganisms ranged from 10^6 to 10^7 CFU/g (18). Senescence and decline of quality of fresh-cut kiwifruit, papaya, and pineapple stored at 4°C to an unsatisfactory level were not correlated with increased microbial growth (132). Levels of bacteria on fresh-processed Lollo Rosso lettuce stored under modified-atmosphere packaging (MAP) conditions exceeded 8 $\log_{10}$ CFU/g after 7 days of storage; however, visual quality evaluation suggested that the product was still acceptable for sale (4).

Unaltered healthy tissues are expected to be a poor substrate for growth of opportunistic microorganisms, whereas damaged or physiologically compromised tissues would deteriorate faster and provide a better substrate for microbial growth. In spinach, large numbers of bacteria were found in areas where the cuticle was broken, resulting in invasion of the internal palisade parenchyma, whereas in healthy unbroken leaves microorganisms were not observed (11). In fact, the growth potential of foodborne pathogens is greater on fresh-cut produce than on intact produce because there are more nutrients available on the cut surface (66). Moreover, fruits and vegetables with cut surfaces are more difficult to decontaminate than whole uncut products. With the use of chlorine as a disinfectant, the levels of *Salmonella* serovar Typhimurium on the surfaces of intact vegetables were reduced 1 to 1.5 $\log_{10}$ CFU/g whereas the reduction was only 0.3 to 0.6 $\log_{10}$ CFU/g for cut vegetables (167).

Despite the lack of correlation of sensory quality with microbial population in some studies, some general conclusions can be made. Microbial qualities of leafy vegetables may be distinguished by visual characteristics, whereas with fruits, high levels of bacteria are more accurately perceived through flavor and aroma. Leafy vegetables contain small amounts of sugars, promoting the proliferation of gram-negative microorganisms, which do not produce nonvolatile compounds or cause the development of soft-rot symptoms. On the other hand, fruits undergo spoilage accompanied by very rapid and predominant growth of lactic acid bacteria and yeast. This growth is often associated with the increased production of secondary metabolites, in particular, lactic, propionic, and acetic acids, which are readily detected as off flavors (94). Some root vegetables with higher contents of simple sugars, such as carrots, will undergo microbial fermentation. The lactic acid bacterium *Leuconostoc mesenteroides* is the primary organism responsible for carrot spoilage (97). In root vegetables such as chicory and potato that have more complex carbohydrate contents, soft-rot symptoms develop (128). Spoilage of stem vegetables such as celery is accompanied by development of acidic flavor under low-oxygen-concentration atmospheres (150).

Sanitation of Equipment and Facilities

Surfaces that are routinely in direct contact with produce are potential sources of contamination. Greater numbers of bacteria are recovered from walls, tables, floors, and equipment after a day of processing (125, 130). Microbial loads on food contact surfaces and equipment vary depending on the microbial quality of the commodity handled and the cleaning and sanitizing programs in use (129). In a recent study, all the evaporator cooling coils tested in chill rooms were contaminated with bacteria (61). In the majority of the cases, the coil-attached bacteria were spoilage bacteria, and in 30% of the facilities, pathogenic bacteria (enterococci, *Staphylococcus aureus,* and *Bacillus cereus*) were found. Dirt accumulates easily in cooling coils and aids in the attachment of bacteria (35). Microbial counts of over 5 $\log_{10}$ CFU/cm were observed at 25% of 891 sites investigated (61). Interestingly, temperature, air velocity, and relative humidity (RH) did not influence the level of bacteria. Shredders were identified as a major source of microorganisms for in-plant contamination. *Pseudomonas* spp. were the predominant contaminants (71). Brushes can also serve as a reservoir for contaminating bacteria (125).

Microbial contamination of air-handling systems can result in contamination of the processing plant. Contamination of air in a meat-processing plant has been shown to influence the shelf lives of stored products (3). It is reasonable to expect that a similar situation could occur in fresh produce-processing facilities; however, the importance of clean air is often overlooked. In rooms where produce was wrapped, the level of microbial contamination was significantly lower than that in rooms where produce was unwrapped (61), underscoring the importance of air decontamination. Several methods are available to clean air, the most common being the use of mechanical filters, particularly the high-efficiency particulate air filters (81). Technology that combines filtration with electrostatic precipitation is widely used to deactivate airborne microbes in meat-processing facilities (87, 182). Electrostatic filtration combined with scanning UV light has also been demonstrated to reduce levels of airborne pathogens (46). However, in most cases a reduction in bacterial population of greater than 2 log CFU/m^3 has not been achieved (46, 81, 87). Technology presently used in medical facilities, a process involving titanium dioxide (TiO_2) in combination with UV light, may be a promising method for enhancing food safety in the fresh-food industry (31, 106). TiO_2 particles catalyze the killing of bacteria by near-UV light due to the generation of free radicals by photoexcited TiO_2 particles (38, 136, 195). The optimal TiO_2 concentration has been found to be 1 mg of TiO_2/ml in a photocatalytic reactor with a UV-B lamp (100). This technique is effective in killing a wide spectrum of organisms, including viruses, bacteria (*Vibrio* sp., *Salmonella,* and *Listeria* sp.), and fungi (100).

Cleaning and Sanitizing Agents

Numerous commercial washing formulations for food-processing facilities are available, including surfactant solutions, combinations of surfactants with organic or mineral acids, and alkaline washes. Chlorine-based sanitizers are the most widely used; however, research suggests that under certain applications other sanitizers are more efficacious. Chlorine dioxide exhibits greater activity against several species of bacteria and yeast than sodium hypochlorite, iodine, quaternary ammonium compounds, glutaradehyde, and phenols on hard surfaces (184). Iodine-based sanitizers have a broad spectrum of activity and are effective against yeast, molds, and vegetative cells of bacteria, but spores are very resistant (133). A 6-$\log_{10}$ reduction in levels of *Listeria* sp. on equipment surfaces was achieved using a 1% povidone-iodine solution (24). Iodophors, common iodine formulations that normally combine elemental iodine with a carrier such as polyvinylpyrrolidone, are less corrosive to metals than chlorine and are not as affected by organic matter. However, iodine products tend to stain equipment and react with sugars and starches of fruits and vegetables, forming a blue-purple color (25). Quaternary ammonium compounds are cationic surfactants commonly used on floors and aluminum equipment. They have good penetrating ability, are stable in the presence of organic matter, and are very effective against molds and *Salmonella* (48, 171). However, these sanitizers have little effect against coliforms and clinical bacteria such as *Listeria* sp. (24, 123).

Other alternatives to chlorine include phenolic-base disinfectants, acid-anionic compound-based compounds, ozone, diethylenetriamine, and peroxyacetic acid, some of which have shown promise (25). Notably, phenolic compounds have shown efficacy in decontaminating surfaces contaminated with norovirus and feline calicivirus (FCV) (78). Chlorine is not particularly effective at inactivating FCV (55).

Combining disinfectants, including alkaline sanitizers, seems to be a good practice to increase bacterial killing. A 5-log reduction in numbers of *L. monocytogenes* on food-processing surfaces occurred when sanitizer application followed cleaning using alkaline solutions, which alone resulted in a <2-log reduction (185). However, the risk of cross-adaptation of bacteria with continuous use of a sanitizer, especially one with extreme pH, is a possibility. Exposure of bacterial cells to extreme pH resulted in cross-protection against environments that would otherwise be lethal (155). Acid-adapted *E. coli* O157:H7 cells are more resistant to heat than unadapted cells (12). *E. coli* O157:H7 also exhibited enhanced resistance to heat after exposure to alkaline cleaners (169). Resistance to other environmental stresses as a result of exposure to acid or alkaline cleaners has not been demonstrated. Cross-adaptation of *L. monocytogenes* to disinfectants with similar mechanisms of action was observed. The bacterium exhibited cross-adaptation to ammonium chloride-based disinfectants and the tertiary amine cationic surfactants. Furthermore, disinfectants with different mechanisms of action were also observed to cause cross-adaptation of *L. monocytogenes*. The results suggest that rotation of disinfecting agents having different mechanisms of action may not be effective. *L. monocytogenes* strains did not adapt to potassium persulfate; however, nonspecific cross-adaptation to other disinfectants was observed (118). Regardless, the strategy of using a combination of sanitizers has been proposed as a possible means to prevent biofilm development (171).

PRODUCE WASH TREATMENTS

Factors Affecting Efficacy of Disinfectants

The desired effect of sanitizing treatments is often not observed, in part because external factors promote rapid and strong attachment of bacteria to the tissue or form barriers that reduce the contact of the sanitizer with the microorganism. Long intervals between the contamination event and washing potentially allow the organism to become more strongly associated with the commodity, reducing the efficiency of the washing treatment. Rapid attachment of pathogenic bacteria to produce surfaces has been demonstrated (113). A reduction of 1 log was obtained by washing produce with water 30 min after bacterial contamination; the reduction was essentially zero when washing was done after 24 h (161). In cantaloupes, *E. coli* and *Salmonella* serovar Stanley populations were reduced by 3 logs when the washing treatment (0.1% sodium hypochlorite or 5% hydrogen peroxide) was applied immediately after inoculation. However, when the antimicrobial agents were applied 72 h postinoculation, the reduction in bacterial populations was less than 1 log (190, 191).

Sanitizing steps may not be successful in certain instances due to attachment of bacteria to inaccessible sites on the produce, including pores, cut surfaces, indentations, and other irregularities (159). Besides the physical barriers, chemical attributes of produce are also involved in the phenomena as attachment to injured tissue is mainly determined by hydrophilic properties of injured surfaces (83). Fresh-cut produce is of special concern due to the topography of the surfaces of these products. Washing with 5% hydrogen peroxide reduced levels of *E. coli* on the surfaces of inoculated apples by nearly 3 logs when the calyxes and stem portions were removed (83). No apparent decrease was noted when the tissues were not removed. In wounded apples, no disinfectant treatment, including washing with acetic acid (5%), hydrogen peroxide (5%), or sodium hypochlorite (0.02%), efficiently reduced growth of *E. coli* O157:H7. In fact, populations of *E. coli* O157:H7 increased from 4 to 7 log CFU/wound after 7 days at room temperature (183). It has been observed that *E. coli* O157:H7 proliferating in cut edges of iceberg lettuce is protected from chlorine wash treatments (166). Evidently, microbes survive in protected sites and can proliferate when the produce is stored at room temperature, even after treatment with sanitizers. Deterioration of a commodity through the action of spoilage organisms can result in an increase in all types of microbes present on the commodity. The growth rate of pathogens on vegetables presenting soft-rot symptoms was twice that on healthy vegetables (197).

Microbes associated with biofilms are protected from the action of sanitizers. Once attached, bacteria can aggregate and become part of a biofilm composed of extracellular polysaccharide matrix that holds the cells together and glues them to the commodity surface (40, 44). Bacteria within a biofilm are more resistant to detachment or inactivation by washing treatments. Human pathogens that are capable of growing in produce, such as *E. coli* O157:H7, *Salmonella* spp., *L. monocytogenes*, and *Pseudomonas*, tend to form biofilms when invading new areas (147, 178). The presence of human pathogens in biofilms on fruits and vegetables and on processing equipment underscores the need for efficacious sanitizers (40, 62). The use of brushes appears to be a good option to remove biofilms; however, unless sanitized continuously, the brushes can easily become a source of microbes (125). The type of surface, the nutrient level, and the organism influence biofilm development and resistance to sanitizers (154). Alternative types of sanitizers, such as products containing lactoferrin that are capable of preventing biofilm formation on beef, may have future application in produce processing (172).

Internalization of human pathogens within produce precludes effective disinfection by washing and sanitizing treatments. Fruits and vegetables with bacteria in internal tissues have been linked to foodborne illness outbreaks (156). Hydrocooling of warm commodities can result in infiltration by the pathogen because the temperature differential and the hydrostatic pressure create a partial vacuum that allows microbes in solution to pass through pores, channels, or punctures into the commodity (36, 204). Internalization also can occur during flowering or fruit development (79, 157). Internalization has been reported for lettuce (166, 176), radishes (91), cucumbers (122, 157) and tomatoes (20, 204). An additional concern with internalization of pathogens is that surviving microorganisms in internal tissue can grow at a faster rate after sanitation of produce. When apples were immersed in chlorine solutions after internalization of *E. coli* O157:H7, the reduction in pathogen levels was 1 to 3 logs, but surviving microorganisms grew at a faster rate and to higher levels than they had presanitation during storage (36).

Pathogens differ in their susceptibilities to sanitizing agents. In lettuce, *L. monocytogenes* grew well after sanitation with chlorine (26, 27), and even better on hydrogen peroxide (10%)-treated endive leaves than in tissue not exposed to a sanitizer (39). *Salmonella* and *Aeromonas* spp. levels seemed to be reduced more readily than those of other pathogens by common disinfectants (67). Gram-negative bacteria are generally more susceptible at low pH than gram-positive bacteria, and weak organic acids have greater inhibition activity than strong acids because they can penetrate and acidify the cell's interior (60). The "hurdle concept," a suitable combination of growth-limiting factors at subinhibitory levels to inhibit growth of microorganisms, is perhaps one of the most successful approaches to reducing risks of contamination (108, 165). For instance, the combination of organic acids with chlorine reduces the solution pH to less than 3, making the chlorine more effective, though it could be more corrosive for equipment (25).

Wash Treatments

Chlorine

Chlorine is the most common sanitizing agent used. Three forms have been approved for use: chlorine gas (Cl_2), calcium hypochlorite ($CaClO_2$), and sodium hypochlorite (NaOCl). The antimicrobial activity of chlorine depends on the amount of free available chlorine in the solution, the pH, the temperature, and the amount of organic matter. Low pHs of internal tissues of produce and high loads of organic matter in the sanitizing solution significantly reduce the effectiveness of chlorine (183).

Dipping of produce into chlorine is broadly used due to its ease of implementation and low cost but appears to have limited benefit, resulting in only a 1- to 2-log reduction in bacterial populations (2, 21, 71, 160, 186). In some cases, chlorine wash reduced the initial bacterial population; however, after a few days no significant differences in bacterial populations on chlorine-washed and water-washed lettuce were evident (26). Counts of *Listeria innocua* and *E. coli* cells in lettuce and coleslaw were significantly reduced, but dipping resulted in enhanced survival and/or growth during extended storage at 8°C (68). Spray application of chlorine to raw produce at retail sites or at the household level may be a suitable, and more convenient, alternative to treatment by dipping or submersion. Reduction in populations (numbers of CFU per square centimeter) of *Salmonella*, *E. coli* O157:H7, *L. monocytogenes*, and total aerobic mesophilic microorganisms on whole apples, tomatoes, and lettuce leaves following spray application of 200 ppm of chlorine (28) was similar to that achieved by dipping produce in chlorinated water (27, 195, 202, 204, 205). Spray application of chlorine at harvest is a practice that is becoming a common approach to reduce bacterial counts in the field (Fig. 1); however, its efficacy is questionable, particularly when mishandling follows the sanitizing treatment.

In a study evaluating the use of sanitizing chemicals for microbial reduction in cantaloupes and honeydew melons, the general order of effectiveness at killing selected pathogens was found to be as follows: chlorine (200 or 2,000 ppm) > acidified sodium chlorite (850 ppm) = Tsunami (a formulation of peroxyacetic acid and hydrogen peroxide; 40 or 80 ppm) > hydrogen peroxide (1%). The first three chemicals reduced populations of *E. coli*

Figure 1 Lettuce for ready-to-eat salads is sprayed with chlorine solution immediately after harvest.

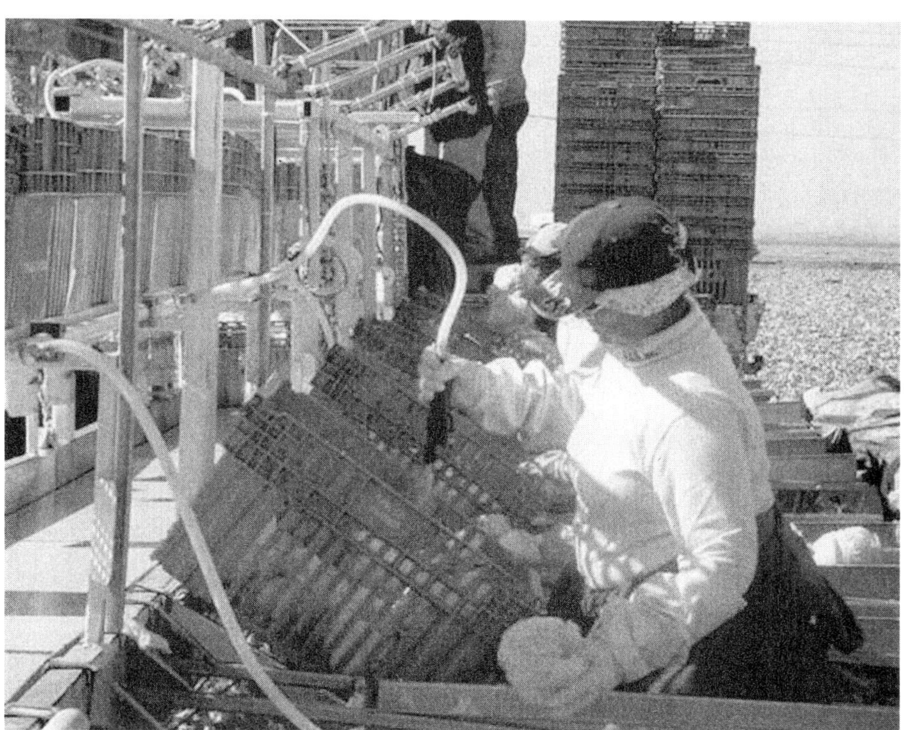

O157:H7 and *Salmonella* by 2.6 to 3.8 $\log_{10}$ CFU/g (139). Interestingly, the authors used an exceptionally high concentration of chlorine (2,000 ppm) because previous results showed very low efficacy of chlorinated water in reducing populations of *E. coli* O157:H7 associated with lettuce (25). Moreover, no significant decrease in *E. coli* O157:H7 levels on fresh-cut lettuce was observed after treatment with chlorine at 20 ppm before or after inoculation at either 50 or 20°C (111). In a separate study (56), exposure for 3 min to a chlorinated solution (47°C) reduced the initial population by 3 log CFU/g.

The addition of a surfactant enhances the lethality of chlorine by increasing surface contact of the sanitizer with the microbe. Dip solutions containing 0.1% of the surfactant Tergitol plus 100 ppm of chlorine resulted in a reduction in *Yersinia enterocolitica* levels of 2.73 $\log_{10}$ CFU/g. Adding 0.5% lactic acid to the same solution produced a reduction of more than 6 $\log_{10}$ CFU/g (60). However, the use of surfactants or other agents in combination with chlorine can adversely affect sensory qualities of produce (25).

Evaluation of chlorine as an enhancer of overall quality of produce has yielded mixed results. In most cases, washes with chlorinated water are not considered effective in preventing decay if fruits and vegetables are already contaminated (57); however, washing of tomatoes with chlorine solutions (250 ppm) significantly reduced the development of decay (19). Calcium hypochlorite solutions have shown good efficacy for some common postharvest spoilage organisms and the plant pathogen *Alternaria alternata* (144). Warm (50°C) chlorine solutions delayed browning and reduced initial populations of several groups of microorganisms naturally occurring on iceberg lettuce but enhanced microbial growth during subsequent storage (108).

Chlorine is considered to be very effective against viruses. In the United States, a chlorine concentration of 5 ppm is recommended to inactivate hepatitis A virus (HAV) and a concentration of 10 ppm is recommended for norovirus (168). A reduction of more than 98% in levels of poliovirus 1 was obtained with 200 ppm of free chlorine (117). However, other viruses such as FCV are resistant to chlorine. A concentration of 1,000 ppm of chlorine is required to ensure complete inactivation of FCV (53).

Chlorine dioxide

Information on the effectiveness of chlorine dioxide (ClO_2) is limited. However, ClO_2 has received attention due to the development of technologies that facilitate shipment and development of simple on-site generation equipment. A positive aspect of the antimicrobial activity of chlorine dioxide is that in the presence of high levels of organic matter, such as those found in immersion dump tanks and flume processing waters, the activity is not substantially diminished and chlorine dioxide does not react with ammonia to form chloramines, as chlorine does (22, 151). In wash water, residual ClO_2 at 1.3 ppm was effective for controlling microbial buildup but surprisingly had little effect on the viability of microorganisms on cucumbers (148).

The method by which a product is exposed to ClO_2 can significantly impact efficacy. Gaseous ClO_2 (3 ppm) was more effective at reducing *L. monocytogenes* populations on both uninjured and injured green pepper surfaces than aqueous ClO_2 (3 ppm) or water washing. The gas treatment produced a reduction of nearly 6 log CFU/g, whereas the aqueous chlorine dioxide produced a 3.5-log reduction and washing with water alone produced a 1.4-log reduction (84). A greater benefit with ClO_2 was observed when washing with water was used to decrease the initial population of bacteria on vegetable surfaces before or after ClO_2 gas treatment. *E. coli* O157:H7 and *L. monocytogenes* inoculated onto strawberries at 7 to 8 log CFU/fruit were eliminated by exposure to 3 ppm of ClO_2 for 10 min (85). A reduction of greater than 3 logs on the calyxes or stem cavities and greater than 5-log reduction on the skin of apples was achieved following treatment with ClO_2 for 30 min at 21°C under 90 to 95% RH (54). ClO_2 gas treatment was effective at killing pathogens on lettuce without adversely affecting visual quality characteristics. Levels of *E. coli* O157:H7, *L. monocytogenes*, and *Salmonella* serovar Typhimurium inoculated onto lettuce leaves decreased by 3.4, 5.0, and 4.3 logs, respectively, after 30 min of exposure to 4.3 ppm of ClO_2. Longer periods of exposure increased the reduction in pathogen populations, especially for *E. coli* O157:H7 (107). Limited information exists on the effect of chlorine dioxide against common viruses such as norovirus and HAV.

Ozone

The effectiveness of ozone gas for the inactivation of microorganisms on the surfaces of fruits and vegetables is influenced by concentration, exposure time, RH, temperature, microbial load, and surface properties of the commodity. Indeed, gas concentration, RH, and time were all significant factors for the inactivation of *E. coli* O157:H7, with gas concentration being the most important factor (82). Similar to that by chlorine, the degree of microbial control exerted by ozone on produce cannot be predicted based on antibacterial activity exhibited in water. Although spores of bacteria are killed very quickly in ozonated water, vegetative cells inoculated into wounds on fruits survived exposure to 1.5 ppm of ozone for 5 min (174). Very little is known about the effect of ozone on virus activity.

Depending on application, the efficacy of ozone can vary considerably. Ozone at 3 ppm was more effective than chlorine dioxide (3 to 5 ppm), chlorinated trisodium phosphate (100 to 200 ppm), and peroxyacetic acid

(80 ppm) at reducing populations of *E. coli* O157:H7 and *L. monocytogenes* in a model system. When applied to produce (apples, lettuce, strawberries, and cantaloupes), ozone and chlorine dioxide (5 ppm) both reduced bacterial populations by 5.6 log (152). However, ozonated water generally reduces bacterial populations in produce by no greater than 3 log CFU/g (42, 101). Ozone was less effective than acidic electrolyzed water at reducing levels of coliforms and aerobic mesophiles in cucumbers (105). A combination of sonication and high-speed stirring has been reported to enhance the efficacy of ozone (101).

One of the benefits of ozone is that it is compatible with bicarbonate salts, which are used to control sour rot and green mold on harvested citrus. Ozone can extend the lives of the bicarbonate solutions by reducing chemical oxygen demand and clarifying the solutions, and it can also kill microbes that contaminate the bicarbonate solutions (174). Excellent results have been observed when grapes have been treated with ozone for controlling postharverst molds and yeasts. The effectiveness of short postharvest exposure of table grapes to ozone for the control of cluster decay is apparently due to both the direct antimicrobial effect of ozone and its ability to elicit stilbene phytoalexins in the grape berries. A 10-min exposure to ozone was as effective as immersion of the fruit in ethanol for reducing the microbial populations on harvested berries (163). The plant defense system elicited by ozone involves the induction of plant response proteins or enzymes capable of detoxifying the active oxygen species and regulation of the phenylpropanoid pathway (98).

Acetic acid, hydrogen peroxide, and peroxyacetic acid

Organic acids, such as acetic acid, have antimicrobial properties and are generally recognized as safe. Reduction of *L. monocytogenes* levels by <1 log on lettuce and cabbage was reported following dipping in 1% acetic acid (202). Higher concentrations can adversely affect the quality of leafy vegetables. Treatment of lettuce with vinegar (1.9% acetic acid) resulted in reduction of poliovirus 1 (117) and *E. coli* (5 $\log_{10}$ CFU/g) levels, but quality was considered unsatisfactory (193). Application of both 5% acetic acid and 80 ppm of peroxyacetic acid was effective in reducing *E. coli* O157:H7 levels on apple surfaces without causing physical injury (198).

The bactericidal activity of hydrogen peroxide is well known. Hydrogen peroxide decomposes rapidly into water and oxygen through the action of catalase, an enzyme commonly found in plants, leaving no residual toxicity. Treatment of lettuce with 2% hydrogen peroxide at 50°C for 60 s was recommended as an effective means to reduce populations of *E. coli* O157:H7, *Salmonella enterica*, and *L. monocytogenes* (115). The combination of lactic acid and hydrogen peroxide reduced levels of the target bacteria by more than 3 logs; however, the sensory quality of the lettuce was negatively affected. Hydrogen peroxide caused browning of apple peels at temperatures greater than 60°C and bleaching of anthocyanins in mechanically damaged berries (160). Solutions of 5% acetic acid and peroxyacetic acid effectively reduced *E. coli* O157:H7 populations by 3 logs (198). A combination of peroxyacetic acid and hydrogen peroxide completely inactivated norovirus and FCV on strawberries and lettuce (78). Tsunami, a formulation of peroxyacetic acid and hydrogen peroxide, reduced levels of poliovirus 1 by 94% (117). In a food service setting, treatment of lettuce with Victory water, a commercial produce wash delivering 60 ppm of peroxyacetic acid and hydrogen peroxide, reduced indigenous microflora by 1.8 logs, whereas washing lettuce with water yielded only a 0.8-log reduction (175).

NOVEL ANTIMICROBIAL TREATMENTS

Modification of conventional treatments can result in increased biocidal activity. An additional 1-log reduction in microbial population on iceberg lettuce was achieved by utilizing ultrasound in combination with a chlorinated rinse solution (167). Ultrasonic fields consist of waves at high amplitude that form cavitation bubbles (170). Cavitation enhances the mechanical removal of attached or entrapped bacteria on the surfaces of fresh produce by displacing or loosening particles through shearing or scrubbing action. Not all microorganisms show the same sensitivity to cavitation (137). Generally, small round cells, gram-positive bacteria, and spores are more resistant than protozoa and gram-negative bacteria (170). It has been shown that there are some exceptions as cavitation has little effect on the viability of *Saccharomyces cerevisiae* (43).

Electrolyzed oxidizing water (EOW) was recently introduced, and it has been determined to be an effective treatment for sanitation of kitchen cutting boards (192) and removal of food pathogens in vitro (99). EOW encompasses at least three antimicrobial properties, low pH (near 2.5), high oxidation-reduction potential (>1,100 mV), and chlorine-based reactants (10 to 90 ppm) (140). Greatest activity appears to be associated with oxidative reduction potential of higher than 848 mV (181). EOW at 15 to 50 ppm of available chlorine was effective in reducing microbial flora on several fresh-cut vegetables (92). EOW washes lasting 3 min significantly decreased populations of *E. coli* O157:H7 and *L. monocytogenes*, by 2.41 and 2.65 $\log_{10}$ CFU per lettuce leaf, respectively (92). The microbial population (total aerobic bacteria, coliform bacteria, *Bacillus cereus*, and psychrotrophic bacteria) in fresh-cut cabbage and lettuce was initially reduced. However, subsequent growth rates were higher than those in untreated vegetables (102). Acidic electrolyzed water at 150 ppm at freezing temperatures reduced *E. coli* O157:H7 populations in lettuce by 2 logs but had little effect on *L. monocytogenes* (104). At 240 ppm, EOW reduced *L. monocytogenes* levels by 1.5 logs but resulted in a burn-like appearance of lettuce (103).

Trisodium phosphate, an alkaline disinfectant generally used for sanitizing meat, showed limited efficacy against pathogens on produce. Trisodium phosphate at 15% completely inactivated *Salmonella* on the surfaces of tomatoes (205). In the same study, only a 2-log reduction in *Salmonella* associated with core tissue was achieved (205). A similar reduction was obtained by washing with water (145). In lettuce, trisodium phosphate solutions that did not damage the sensory properties resulted in essentially no reduction of *L. monocytogenes* populations (178, 202).

Calcinated calcium may be a promising agent to control pathogenic microorganisms on fresh produce. A 200-ppm chlorine solution reduced *L. monocytogenes* numbers by 2.27 $\log_{10}$ CFU per tomato; however, treatment with calcinated calcium resulted in a reduction of 7.59 $\log_{10}$ CFU per tomato (17). Calcinated calcium at 1 and 2% reduced *Salmonella* populations on alfalfa seeds by 3.2 logs (196) and 1.5 logs (69), respectively, but the germination rate was negatively affected. Results of these studies are encouraging and justify further research.

Another produce sanitizer becoming very common in Europe is potassium permanganate. In six college food establishments, the effect of washing lettuce with sodium hypochlorite and potassium permanganate solution was evaluated. The incoming lettuce contained counts of nearly 7 log CFU of mesophilic bacteria/g. Washing lettuce for 2 min in sodium hypochlorite at 70 ppm or potassium permanganate at 25 ppm reduced populations of aerobic bacteria and total coliforms by nearly 2 logs (179). One convenient feature of using potassium permanganate is that the solution produces a pink color, which helps to indicate whether the rinsing procedure has been correctly accomplished.

Cinnamaldehyde, a natural compound used for flavor, exhibits antimicrobial activity (95). Treatment of tomatoes with an aqueous solution of 13 mM cinnamaldehyde reduced the number of bacteria and fungi by 1 $\log_{10}$ CFU/g within 10 min. On tomatoes treated for 30 min with cinnamaldehyde, visible mold growth was delayed by 7 days during storage under modified atmosphere conditions at 18°C (173). When cinnamaldehyde was mixed with carvacrol, the activity on low-pH produce was enhanced. Carvacrol and cinnamaldehyde at 0.15 ppm were effective at reducing bacteria on kiwifruit but not bacteria associated with honeydew melons (153). The specificity of these antimicrobial agents highlights the need for research evaluating mixed treatments in order to properly design a multitreatment system.

The antimicrobial activity of essential oils from spices and culinary herbs appears to be associated in most cases with phenolic compounds (37, 40). Purified compounds derived from essential oils such as carvacrol, eugenol, linalol, and thymol inhibit various microorganisms (14). Among seven individual oil components tested against 25 bacterial strains, the component with the widest spectrum of activity was thymol, followed by carvacrol (52). The level of *Shigella* spp. on lettuce leaves treated by immersion for 2 min in a 0.1% solution of thymol and carvacrol decreased to below the detection limit (14). In the same study, aerobic bacteria decreased from 6 log CFU/g to 2 log CFU/g. A number of nontraditional disinfecting agents are being evaluated, especially those found in nature (25). For example, alcohols and aldehydes derived from green leaves exhibited strong bacteriostatic activity against *Staphylococcus aureus* IFO 12732, *E. coli* O157:H7, and *Salmonella* serovar Enteritidis (127).

TECHNOLOGY APPLICATION

The methods and equipment used to wash produce in packing facilities have come under scrutiny. In some cases, sanitizing agents tested in the laboratory yielded a reduction of 2 to 3 logs, whereas in commercial facilities a reduction of no more than 1 log in bacterial numbers has been reported (10). This lack of efficacy has been attributed in part to poorly designed equipment that results in insufficient exposure of the contaminated commodity surfaces. This suggests that poorly designed equipment can negatively impact microbial reduction strategies.

Various types of washers are used for produce, including brush washers, reel washers, pressure washers, hydro air agitation wash tanks, and immersion pipeline washers (158). Systems that apply the disinfectant during the brushing operation seem to be ideal, preventing bacterial buildup on the brushes (125). Systems designed to rub or brush the commodity surface enhance the removal of pathogens, but the equipment can also serve as a source of contamination if not sanitized properly (117).

Typically, sanitizers are applied by spraying or dipping the produce; however, novel means of applying sanitizing agents are being evaluated. Vapor-phase treatment is a promising technology that seems to permit sanitizing agents to efficiently reach microbes in hard-to-access sites. Acetic acid applied in vapor phase reduced the microbial populations in cabbage, mung bean seeds, and grapes (49, 158). *E. coli* populations in inoculated apples were reduced by 3 logs with acetic acid in vapor phase, but the treatment also resulted in browning of tissue (161). Gaseous hydrogen peroxide yielded a 2-log reduction in *E. coli* levels, whereas the same treatment with chlorine dioxide gas at 68°F resulted in a reduction of more than 4 $\log_{10}$ CFU/g in the bacterial population without affecting the quality of apples (162). While water washing achieved only a 1.5-log reduction, treating green peppers with 0.62 and 1.24 ppm of ClO_2 in a gaseous phase resulted in 3.03- and 6.45-log reductions in *E. coli* O157:H7 populations, respectively (83). In general, gaseous ClO_2 has yielded encouraging results under laboratory conditions, but very little has been reported on its application in commercial settings.

Vacuum infiltration appears to be another promising system to maximize contact of sanitizers with microorganisms on produce. This technique was first utilized to increase uptake of browning inhibitors by fresh-cut apples (159) and uptake of calcium by intact apples (146). It has been tested with 5% hydrogen peroxide solution on inoculated apples, yielding a 4- to 5-log reduction in *E. coli* populations without any noticeable damage to the post-harvest quality of the fruit (158). The use of vacuum infiltration of hydrogen peroxide has consistently yielded high effectiveness in several laboratory trials without drawbacks in visual or taste quality (161).

Surface pasteurization with steam, hot water, or air is an alternative presently used in commercial settings. It seems very effective with agents such as hydrogen peroxide, particularly for produce with hard surfaces (158), because the steam process can damage produce with delicate tissue. Exposure of *E. coli*- and *Salmonella*-contaminated cantaloupes to 5% hydrogen peroxide solutions at 80°C for 3 min resulted in a 4-log reduction in levels of the pathogens with no signs of damage after storage at 4°C for 26 days (158). Surface pasteurization of *Salmonella*- or *E. coli*-contaminated melons at 47°C for 3 min 24 h after inoculation reduced pathogen populations by 5 logs and slowed senescence (9).

Heat treatment virtually eliminates decay on fruit contaminated prior to heating; however, it has little effect when contamination occurs after heating. When heat-treated apples were purposely injured and inoculated with pathogens, the postsanitation growth of microorganisms was greater (109). This was probably due to the destruction of enzymes that are used for defense against pathogen invasion. Combining heat treatment with sanitizing treatment may provide a synergistic effect; however, this possibility requires validation for each combination and application.

Few reports address the influence of heat treatments on the shelf life of fresh-cut produce. In general, for a commodity such as fresh-cut lettuce, most studies have shown that warm aqueous treatments tend to diminish the overall quality of the product during long-term storage, especially when the treatment lasts for more than 3 min. Warm water (45 to 50°C) applied for less than 1 min reduced the microbial population by 1 log CFU/g while reducing browning of tissue during storage (121, 134). Novel heating methods such as radio frequency treatment appear to have some benefits over distinctive traditional heat methods. In radio frequency treatment, the heat is transferred from the inside of the product to the outside. The technology would be relatively simple to adapt to a processing line. However, when the treatment was applied to carrot sticks, reduction in bacterial numbers was no better than that by hot water treatment, which reduced bacterial counts from 2×10^5 log CFU/g to less than 10 CFU/g (138).

Nonaqueous Treatments

Irradiation

Irradiation has been shown to be effective at decreasing microbial counts on intact and fresh-cut produce. This treatment is recommended for use on fruits and vegetables at a maximum level of 1.0 kGy. A 1.0-kGy dose effectively reduced mesophilic bacteria in Mexican salads (90). A dose of 0.19 kGy resulted in significant decreases in populations of aerobic bacteria and yeasts in cut iceberg lettuce (80). However, doses needed to accomplish microbial reduction through radiation may lead to changes in pectic substances, resulting in softening of the plant tissue (47, 89). An exception to this has been reported for celery. Irradiating celery contaminated with *L. monocytogenes* and *E. coli* at 1.0 kGy eliminated both pathogens and extended the shelf life from 22 days (control) to 29 days (143). The color, texture, and aroma of the celery were not affected (143).

The effectiveness of gamma irradiation for viruses is dependent on various factors such as the size of the virus, the type of product, and the temperature of exposure (45). Gamma irradiation between 2.7 and 3.0 kGy is required to achieve more than 90% deactivation of HAV on fruits and vegetables (30). The 1-kGy dose approved for food irradiation in the United States may not be sufficient to inactivate viruses such as HAV since a 1-kGy dose results in only 90% deactivation of HAV in lettuce and strawberries (29).

UV light

UV light at a wavelength of 200 to 280 nm appears to be a good sanitation alternative for certain produce applications. UV-C light offers several advantages: (i) no residue remains after treatment, (ii) it requires no subsequent removal of moisture, and (iii) it does not involve complex safety equipment. Bacterial spores are generally more resistant to UV-C treatment than bacteria in a vegetative state. Cells in exponential growth are commonly less resistant than cells in stationary phase. UV-C light also provides secondary benefits for the quality of produce, including reduction of postharvest decay of various fruits and vegetables (5, 75, 180). The reduction in postharvest diseases and delay of decay in commodities exposed to low-level UV-C radiation might be related to the increase in decay resistance of tissues due to the accumulation of antifungal compounds such as polyamines (59). Moreover, UV-C treatment reduces breakdown and chilling injury of peaches (74).

UV-C treatments were effective at reducing growth of psychrotrophic bacteria, coliforms, yeast, and molds (5). Growth of lactic acid bacteria on lettuce was stimulated following UV-C treatment, probably due to higher relative resistance and lack of competing microbes. UV light at a dose exceeding 9 mW/cm^2 resulted in a 2-log reduction in microbial populations on lettuce and tomatoes and a 3-log reduction in microbial populations on apples (200).

Disadvantages of UV-C treatment include poor penetration and potential negative effects on product quality at high doses. UV-C radiation increased the respiration rate of fresh-cut lettuce, indicating that stress was posed to the tissue with the radiation treatment. Greater reductions in bacterial populations are observed in produce with smoother surfaces since occlusion of the light path with water or with tissue negatively impacts efficacy (16). UV-C treatment reduces virus infectivity but does not completely inactivate viruses such as noroviruses and FCV (131).

Modified atmosphere

MAP involves the packaging of food under an atmosphere that differs from the normal air composition. Concentrations of nitrogen, oxygen, and carbon dioxide are adjusted depending on the type of product and the desired shelf life. In most cases, environments with elevated CO_2 and/or reduced O_2 levels extend the shelf life of produce by inhibiting chemical, enzymatic, and microbial spoilage. The atmosphere conditions can selectively inhibit the growth of gram-negative bacteria such as *Pseudomonas* spp., which under aerobic conditions typically grow rapidly and produce the off flavors and odors associated with the spoilage of many foods (67). Microorganisms such as lactic acid bacteria that are not usually affected by high CO_2 levels tend to predominate when MAP is used, actually prolonging shelf life by producing compounds that can inhibit spoilage bacteria. Longer shelf life associated with high CO_2 levels has been observed in various applications (76, 142). A possible disadvantage is that levels of CO_2 necessary to inhibit *Pseudomonas* spp. and enhance growth of lactic acid bacteria are normally above 10%, often high enough to affect the visual quality of produce. Similarly, very low O_2 concentrations can suppress the growth of many spoilage bacteria but they can also induce fermentation of commodities such as fresh-cut watermelon

(65), citrus (76), and sweet potatoes (41). Although MAP can potentially inhibit the growth of bacteria and prevent fungal spoilage, it has no effect on the enteric viruses, including HAV, under common refrigeration temperatures (30).

The growing popularity of minimally processed fresh produce has resulted in increased use of plastic films and subsequently in the use of MAP. It has been shown that high survival rates of pathogenic bacteria such as *Shigella flexneri* and *Shigella sonnei* are feasible in common MAP conditions. The growth of these two microorganisms continues after processing but is variable depending upon the type of vegetable mix. For example, in mixed lettuce products, the numbers of *Shigella* sp. were over 5 $\log_{10}$ CFU/g after 7 days of storage, whereas in carrots counts of 2 $\log_{10}$ CFU/g were found (13). High concentrations of CO_2 have also resulted in lower bacterial counts on mango cubes; variation in effectiveness between varieties was noted (142). MAP at 100% nitrogen did not significantly affect the growth of microbial populations in cut vegetables (102). Moderate vacuum packaging, which normally involves product packaged under 40 kPa of atmospheric pressure, may substantially reduce levels of pathogens such as *L. monocytogenes*, *Salmonella* serovar Typhimurium, *Yersinia enterocolitica*, and *Bacillus cereus* (77). Atmospheres with O_2 levels higher than 70 kPa may be most appropriate for maintaining sensory quality and safety. Superatmospheric O_2 levels inhibit growth of lactic acid bacteria and *Enterobacteriaceae;* however, *L. monocytogenes* is not affected (6). In general, certain MAP conditions may be used to further suppress growth of some microorganisms; however, extending freshness of produce could potentially provide the microorganisms more time to grow.

Edible coatings

The application of biodegradable edible coatings, such as sucrose polyesters of fatty acids, proteins, cellulose derivatives, starch, and other polysaccharides, onto produce is an ongoing topic of research for many scientists. Under certain conditions of RH and temperature, edible coatings provide a good barrier to oxygen and carbon dioxide transmission that can lower respiration and senescence metabolism. Biodegradable coatings can potentially extend shelf life and, with the inclusion of an antimicrobial, reduce the potential growth of pathogens (15). Coatings with sorbitol reduced microbial counts, and the addition of potassium sorbate extended the shelf life of strawberries (70). Shellac coatings with pH 9, ethanol at 12%, and the preservative paraben reduced the population of *E. coli* and *Klebsiella pneumoniae* from 5 log CFU/cm^2 to 1.4 log CFU/cm^2 in citrus (120).

Chitosan, a deacetylated form of chitin obtained from crustacean shells or fungi, degrades phospholipid components of cell membranes (116). The antimicrobial activity of chitosan depends on factors such as the type of chitosan (deacetylation degree and molecular weight), the pH of the medium, the temperature, and the presence of food components. Chitosan-lactate polymers (ranging from 0.5 to 1.2 MDa) inhibited the growth of *Saccharomyces bayanus* and *Saccharomyces unisporus* in fermented vegetables (164). In the same study, native chitosan showed no inhibition and it was shown that the chitosan-lactate activity directly promoted cell wall degradation. At pH 6.0, the antimicrobial activity of chitosan was significantly lower than that at pH 4.0. Furthermore, the activity was completely abolished when 10% whey protein was added. A 2-log reduction in bacterial populations on strawberries treated with chitosan and lactic acid-sodium lactate solution occurred during 12 days of storage at 7°C (50). In contrast, no reduction in bacterial populations was noted for lettuce treated with the same solution. Chitosan films enriched with oregano oil reduced *L. monocytogenes* populations by 4 logs, whereas pure chitosan resulted in only a 2-log reduction (206). Oregano oil may impart an unacceptable flavor to the treated product, but perhaps other plant extracts with similar biocidal activities could be used in combination with chitosan on fruits and vegetables.

POSTPROCESSING STRATEGIES AND TECHNOLOGY

A new potential treatment strategy focuses on antimicrobial substances produced by selected vegetables and microorganisms (26). For example, populations of *L. monocytogenes* decreased upon contact with whole and shredded raw carrots. Small populations of *L. monocytogenes* detected on whole carrots immediately after dipping were essentially nondetectable after 7 days of storage at 5 or 15°C, and the carrots spoiled before *L. monocytogenes* grew. Freeze-dried spinach powder had an inhibitory effect on the growth of *L. monocytogenes;* however, the effect was markedly decreased when the native microorganisms were almost eliminated by heating or irradiation (11). Similarly, *L. monocytogenes* populations declined rapidly on tomatoes (141).

Efficient inhibition of the growth of pathogens by antibacterial-producing lactic acid bacteria on fresh-cut vegetables may be achieved by strains that grow and secrete antimicrobial compounds under refrigeration conditions (23). Lactic acid, bacteriocins, and various nonproteinaceous low-molecular-weight compounds are produced by lactic acid bacteria (73). Lactic acid bacteria isolated from biofilms formed in floor drains of a processing facility reduced *L. monocytogenes* growth by over 5 log CFU/cm^2 (203). Nisin, a broad-spectrum pore-forming bacteriocin produced by lactic acid bacteria, is active against many gram-positive bacteria, including *L. monocytogenes* (88). Nisin is especially active at the lower pH typical of many fruits and some vegetables (187) and can be added to packaging films for antibacterial purposes. Nisin applied by spray application reduced *L. monocytogenes* populations by 3.2 logs on honeydew slices and by 2.0 logs on apple slices compared to the control populations (109, 110).

THE FUTURE: DEVELOPMENT OF POSTHARVEST DISINFECTING SYSTEMS

The information that is available on produce decontamination underscores the lack of strong sanitizer alternatives to the conventional systems presently used (e.g., chlorine spraying and dipping). The popular misconception that postharvest treatments can function as killing steps for contaminated produce or as significant reducers of indigenous microbial populations was discussed in this chapter. Although experimentation under laboratory conditions has shown significant reduction of microorganisms, even to undetectable levels, the reports are not always consistent. The best reduction under commercial conditions is generally no greater than 2 to 3 logs at concentrations that are not detrimental to sensory quality but in most cases are unsuitable for application at the household level.

It is of concern that common sanitizers such as hydrogen peroxide and sodium hypochlorite are ineffective when treated produce has punctures or wounds. The antimicrobial agents do not penetrate to inaccessible sites, which suggests that removing injured produce from packing and processing facilities is a wise policy to incorporate in food safety programs. Clearly, if fruits and vegetables are not handled properly after harvest, sanitizing will do little to reduce high bacterial loads. If produce is contaminated postsanitization, the microbial levels may increase to pretreatment levels. Systems that maintain fruits and vegetables in wet conditions after the application of a sanitizer provide the surviving microorganisms ideal conditions to grow. Other conditions promoting rapid postsanitation growth of microbes include high irradiation dosages, rough brushing, and high levels of shock and vibration during transportation since these weaken plant tissue, releasing nutrients for microbial growth.

The safety of fresh-cut fruits and vegetables is a concern because cut surfaces provide sites for bacterial attachment and growth. Technologies applied to surface decontamination and sanitation of fresh-cut produce need to be validated by using surface-injured fruits and vegetables. Differences in surface characteristics of the commodity, the type and physiological state of microbial cells, and environmental stress conditions interact to influence the activity and efficacy of sanitizers. There is a lack of information on the survival of specific viruses on fresh produce, and more research needs to focus on the effectiveness of present washing and decontamination processes for the removal of viruses.

Information on the relationship between postharvest biology of produce and microbial growth, particularly the growth of foodborne pathogens, is limited. More research is needed to understand the effect of biostimulants on microbial growth on plant tissue. It is critical to focus on treatments and handling steps that support the natural resistance of the product to microbial growth. UV-C and ozone treatments appear to be two good examples of this type of treatment, but it is unclear when and how these need to be applied to obtain maximum benefits.

In developing new or improved washing and sanitizing treatments for fruits and vegetables, it is essential to understand the compatibility of the sanitizers with the application systems. The information captured in this chapter suggests that the efficacy of sanitizers depends on the type of produce, the microorganism present, and the application system. Utilization of sequential treatments that have different modes of actions seems to be an interesting approach, providing the costs can be kept at affordable levels. Research demonstrates that a hurdle approach increases reduction in microbial populations by at least twice that obtained with only a single decontamination step. Many of the practices and technologies discussed in this chapter could be used in combination to obtain a hurdle approach that effectively reduces the microbial load on intact and fresh-cut produce. Furthermore, it is imperative to keep in mind that the produce industry cannot rely only on sanitizers to maintain low levels of bacteria on fruits and vegetables. A systematic effort to prevent contamination during all postharvest operations is essential and undoubtedly requires an integrated approach.

REFERENCES

1. **Ackers, M., R. Pagaduan, G. Hart, K. D. Greene, S. Abbott, E. Mintz, and R. V. Tauxe.** 1997. Cholera and sliced fruit: probably secondary transmission from an asymptomatic carrier in the United States. *Int. J. Infect. Dis.* **1**:212–214.
2. **Adams, M. R., A. D. Hartley, and L. J. Cox.** 1989. Factors affecting the efficacy of washing procedures used in the production of prepared salads. *Food Microbiol.* **6**:68–77.
3. **Al-Dagal, M., O. Mo, D. Y. C. Fung, and C. Kastner.** 1992. A case study of the influence of microbial quality of air on product shelf life in a meat processing plant. *Dairy Food Environ. Sanitation* **12**:69–70.
4. **Allende, A., E. Aguayo, and F. Artes.** 2004. Microbial and sensory quality of commercial fresh processed red lettuce throughout the production chain and shelf life. *Int. J. Food Microbiol.* **91**:109–117.
5. **Allende, A., and F. Artés.** 2003. UV-C radiation as a novel technique for keeping quality of fresh processed "Lollo Rosso" lettuce. *Food Res. Int.* **36**:739–746.

6. **Allende, A., L. Jacxsens, F. Devlieghere, J. Debevere, and F. Artes.** 2004. Effect of super-atmospheric oxygen packaging on sensorial quality, spoilage, and *Listeria monocytogenes* and *Aeromonas caviae* growth in fresh processed mixed salads. *Int. Assoc. Food Prot.* **65:** 1565–1573.
7. **Anderson, J. B., T. A. Shuster, K. E. Hansen, A. S. Levy, and A. Volk.** 2004. A camera's view of consumer food-handling behaviors. *J. Am. Diet. Assoc.* **104:**186–191.
8. **Angelillo, I. F., M. R. Foresta, C. Scozzafava, and M. Pavia.** 2001. Consumers and food-borne diseases: knowledge, attitudes and reported behavior in one region of Italy. *Int. J. Food Microbiol.* **64:**161–166.
9. **Annous, B. A., A. M. Burke, and J. E. Sites.** 2004. Surface pasteurization of whole cantaloupes inoculated with *Salmonella poona* or *Escherichia coli*. *J. Food Prot.* **67:**1876–1885.
10. **Annous, B. A., G. M. Sapers, A. M. Mattrazzo, and D. C. R. Riordan.** 2001. Efficacy of washing with commercial flatbed brush washer, using conventional and experimental washing agents, in reducing populations of *Escherichia coli* on artificially inoculated apples. *J. Food Prot.* **64:**159–163.
11. **Babic, I., and A. E. Watada.** 1996. Microbial populations of fresh-cut spinach leaves affected by controlled atmospheres. *Postharvest Biol. Technol.* **9:**187–193.
12. **Bacon, R. T., J. N. Sofos, P. A. Kendall, K. E. Belk, and G. C. Smith.** 2003. Comparative analysis of acid resistance between susceptible and multi-antimicrobial-resistant *Salmonella* strains cultured under stationary-phase acid tolerance-inducing and noninducing conditions. *J. Food Prot.* **66:**732–740.
13. **Bagamboula, C. F., M. Uyttendaele, and J. Debevere.** 2002. Growth and survival of *Shigella sonnei* and *S. flexneri* in minimally processed vegetables packed under equilibrium modified atmosphere and stored at 7°C and 12°C. *Food Microbiol.* **19:**529–536.
14. **Bagamboula, C. F., M. Uyttendaele, and J. Debevere.** 2003. Inhibitory effect of thyme and basil essential oils, carvacrol, thymol, estragol, linalool and *p*-cymene towards *Shigella sonnei* and *Sh. flexneri*. *Food Microbiol.* **21:**33–42.
15. **Baldwin, E. A., M. O. Nisperos-Carriedo, and R. A. Baker.** 1995. Use of edible coatings to preserve quality of lightly (and slightly) processed products. *Crit. Rev. Food Sci. Nutr.* **35:**509–524.
16. **Bank, H. L., J. John, M. K. Schmehl, and R. J. Dratch.** 1990. Bactericidal effectiveness of modulated UV light. *Appl. Environ. Microbiol.* **56:**3888–3889.
17. **Bari, M. L., Y. Inatsu, S. Kawasaki, E. Nazuka, and K. Isshiki.** 2002. Calcinated calcium killing of *Escherichia coli* O157:H7, *Salmonella,* and *Listeria monocytogenes* on the surface of tomatoes. *J. Food Prot.* **65:**1706–1711.
18. **Barriga, M. I., G. Trachy, C. Willmot, and R. E. Simard.** 1991. Microbial changes in shredded iceberg lettuce stored under controlled atmospheres. *J. Food Sci.* **56:**1586–1599.
19. **Bartz, J. A.** 1999. Washing fresh fruits and vegetables: lessons from treatment of tomatoes and potatoes with water. *Dairy Food Environ. Sanitation* **19:**853–864.
20. **Bartz, J. A., and R. K. Schowalter.** 1981. Infiltration of tomatoes by bacteria in aqueous suspension. *Phytopathology* **71:**515–518.
21. **Behrsing, J., S. Winkler, P. Franz, and R. Premier.** 2000. Efficacy of chlorine for inactivation of *Escherichia coli* on vegetables. *Postharvest Biol. Technol.* **19:**187–192.
22. **Benarde, M. A., B. M. Israel, V. P. Olivieri, and M. L. Granstorm.** 1965. Efficiency of chlorine dioxide as a bactericide. *Appl. Microbiol.* **13:**776–780.

23. **Bennik, M. H. J., W. V. Overbeck, E. J. Smid, and L. G. M. Gorris.** 1999. Biopreservation in modified atmosphere stored mungbean sprouts: the use of vegetable-associated bacteriocinogenic lactic acid bacteria to control the growth of *Listeria monocytogenes*. *Lett. Appl. Microbiol.* **28**:226–232.
24. **Best, M., M. E. Kennedy, and F. Coates.** 1990. Efficacy of a variety of disinfectants against *Listeria* spp. *Appl. Environ. Microbiol.* **56**:377–380.
25. **Beuchat, L. R.** 1998. *Food Safety Issues.* World Health Organization, Geneva, Switzerland.
26. **Beuchat, L. R., and R. E. Brackett.** 1990. Inhibitory effect of raw carrots on *Listeria monocytogenes*. *Appl. Environ. Microbiol.* **56**:1734–1742.
27. **Beuchat, L. R., and R. E. Brackett.** 1990. Survival and growth of *Listeria monocytogenes* on lettuce as influenced by shredding, chlorine treatment, modified atmosphere packaging and temperature. *J. Food Sci.* **55**:755–758, 870.
28. **Beuchat, L. R., B. V. Nail, B. B. Adler, and M. R. S. Clavero.** 1998. Efficacy of spray application of chlorinated water in killing pathogenic bacteria on raw apples, tomatoes, and lettuce. *J. Food Prot.* **61**:1305–1311.
29. **Bidawid, S., J. M. Farber, and S. A. Sattar.** 2000. Inactivation of hepatitis A virus (HAV) in fruits and vegetables by gamma irradiation. *Int. J. Food Microbiol.* **57**:91–97.
30. **Bidawid, S., J. M. Farber, and S. A. Sattar.** 2001. Survival of hepatitis A virus on modified atmosphere-packaged (MAP) lettuce. *Food Microbiol.* **18**:95–102.
31. **Blake, D. M.** 1995. *Bibliography of Work on the Photocatalytic Removal of Hazardous Compounds from Water and Air, Update Number 1, to June 1995.* National Renewable Energy Laboratory, U.S. Department of Energy, Washington, D.C.
32. **Boyette, M. D., D. F. Ritchie, S. J. Carballo, S. M. Blankenship, and D. C. Sanders.** 1993. Chlorination and postharvest disease control. *HortTechnology* **3**:395–400.
33. **Brackett, R. E.** 1999. Incidence, contributing factors, and control of bacterial pathogens in produce. *Postharvest Biol. Technol.* **15**:305–311.
34. **Brandl, M. T., and R. E. Mandrell.** 2002. Fitness of *Salmonella enterica* serovar Thompson in the cilantro phyllosphere. *Appl. Environ. Microbiol.* **68**:3614–3621.
35. **Braun, R. H.** 1986. Problem and solution to plugging of a finned-tube cooling coil in an air handler. *ASHRAE Trans. 1B* **92**:385–387.
36. **Buchanan, R. L., S. G. Edelson, R. L. Miller, and G. M. Sapers.** 1999. Contamination of intact apples after immersion in an aqueous environment containing *Escherichia coli* O157:H7. *J. Food Prot.* **62**:444–450.
37. **Burt, S.** 2004. Essential oils: their antibacterial properties and potential applications in foods—a review. *Int. J. Food Microbiol.* **94**:223–253.
38. **Cai, R., Y. Kubota, T. Shuin, H. Sakai, K. Hashimoto, and A. Fujishima.** 1992. Induction of cytotoxicity by photoexcited TiO_2 particles. *Cancer Res.* **52**:2346–2348.
39. **Carlin, F., C. Nguyen-The, A. Abreu da Silva, and C. Cochet.** 1996. Effects of carbon dioxide on the fate of *Listeria moncytogenes,* of aerobic bacteria and on the development of spoilage in minimally processed fresh endive. *Int. J. Food Microbiol.* **32**:159–172.
40. **Carmichael, I., I. S. Harper, M. J. Coventry, P. W. J. Taylor, J. Wan, and M. W. Hickey.** 1999. Bacterial colonization and biofilm development on minimally processed vegetables. *J. Appl. Microbiol.* **85**:45S–51S.
41. **Chang, L. A., and S. J. Kays.** 1981. Effect of low oxygen on sweet potato roots during storage. *J. Am. Soc. Hortic. Sci.* **106**:481–483.

42. **Cherry, J. P.** 1999. Improving the safety of fresh produce with antimicrobials. *Food Technol.* **53**:54–58.
43. **Ciccolini, L., P. Taillander, A. M. Wilhelm, H. Delmas, and P. Strehaiano.** 1997. Low frequency thermo-ultrasonication of *Saccharomyces cerevisiae:* effects of temperature and ultrasonic power. *Chem. Eng. J.* **65**:145–149.
44. **Costerton, J. W.** 1995.Overview of microbial biofilms. *J. Ind. Microbiol.* **15**:137–140.
45. **Croci, L., D. D. Medici, C. Scalfaro, A. Fiore, and L. Toti.** 2002. The survival of hepatitis A virus in fresh produce. *Int. J. Food Microbiol.* **73**:29–34.
46. **Cundith, C. J., C. R. Kerth, W. R. Jones, T. A. McCaskey, and D. L. Kuhlers.** 2002. Air-cleaning system effectiveness for control of airborne microbes in a meat-processing plant. *J. Food Sci.* **67**:1170–1174.
47. **d'Armour, J., C. Gosselin, J. Arul, F. Castaigne, and C. Willemont.** 1993. Gamma-radiation affects cell wall composition of strawberries. *J. Food Sci.* **58**:182–185.
48. **Davison, S., C. E. Benson, and R. J. Eckroade.** 1996. Evaluation of disinfectants against *Salmonella enteriditis. Avian Dis.* **40**:272–277.
49. **Delaquis, P. J., P. L. Sholber, and K. Stanich.** 1999. Disinfection of mung bean seed with vaporized acetic acid. *J. Food Prot.* **62**:953–957.
50. **Devlieghere, F., A. Vermeulen, and J. Debevere.** 2004. Chitosan: antimicrobial activity, interactions with food components and applicability as a coating on fruit and vegetables. *Food Microbiol.* **21**:703–714.
51. **Dong, H., T. P. Delaney, D. W. Bauer, and S. V. Beer.** 1999. Harpin induces resistance in *Arabidopsis* through systemic acquired resistance pathway mediated by salicylic acid and the NIM1 gene. *Plant J.* **20**:207–215.
52. **Dorman, H. J. D., and S. G. Deans.** 2000. Antimicrobial agents from plants: antibacterial activity of plant volatile oils. *J. Appl. Microbiol.* **88**:308–316.
53. **Doultree, J. C., J. D. Druce, C. J. Birch, D. S. Bowden, and J. A. Marshall.** 1999. Inactivation of feline calicivirus, a Norwalk virus surrogate. *J. Hosp. Infect.* **41**:51–57.
54. **Du, J., Y. Han, and R. H. Linton.** 2003. Efficacy of chlorine dioxide gas in reducing *Escherichia coli* O157:H7 on apple surfaces. *Food Microbiol.* **20**:583–591.
55. **Duizer, E., P. Bijkerk, B. Rockx, A. de Groot, F. Twisk, and M. Koopmans.** 2004. Inactivation of caliciviruses. *Appl. Environ. Microbiol.* **70**:4538–4543.
56. **Dychdala, G. R.** 1999. Chlorine and chlorine compounds, p. 131–151. *In* S. S. Block (ed.), *Disinfection, Sterilization and Preservation.* Lea and Febiger, Philadelphia, Pa.
57. **Eckert, J. W., and N. F. Sommer.** 1967. Control of disease of fruits and vegetables by post-harvest treatments. *Annu. Rev. Phytopathol.* **5**:391–432.
58. **Edgar, R., and K. E. Aidoo.** 2001. Microflora of blanched minimally processed fresh vegetables as components of commercial chilled ready-to-use meals. *Int. J. Food Sci. Technol.* **36**:107–110.
59. **Ekran, M., C. Y. Wang, and D. T. Krizek.** 2001. UV-C radiation reduces microbial populations and deterioration in *Cucurbia pepo* fruit tissue. *Environ. Exp. Bot.* **45**:1–9.
60. **Escudero, E. M., L. Velazquez, M. S. Di Genaro, and A. M. S. de Guzman.** 1999. Effectiveness of various disinfectants in the elimination of *Yersinia enterocolitica* on fresh lettuce. *J. Food Prot.* **62**:665–669.
61. **Evans, J. A., S. L. Russel, C. James, and J. E. L. Corry.** 2004. Microbial contamination of food refrigeration equipment. *J. Food Eng.* **62**:225–232.

62. **Fett, W. F.** 2000. Naturally occurring biofilms on alfalfa and other types of sprouts. *J. Food Prot.* **63**:625–632.
63. **Fisher, T. L., and D. A. Golden. 1998.** Fate of *Escherichia coli* O157:H in ground apples used in cider production. *J. Food Prot.* **61**:1372–1374.
64. **Fonseca, J. M.** 2005. New technologies currently being used for produce safety. *Proc. N.J. Vegetable Growers Assoc.* **50**:79–82.
65. **Fonseca, J. M., J. W. Rushing, and R. F. Testin.** 2004. The anaerobic compensation point of fresh-cut watermelon and postprocess implications. *HortScience* **39**:562–566.
66. **Francis, G. A., and D. O'Beirne.** 1998. Effects of the indigenous microflora of minimally processed lettuce on the survival and growth of *L. monocytogenes. Int. J. Food Sci. Technol.* **33**:477–488.
67. **Francis, G. A., C. Thomas, and D. O'Beirne.** 1999. The microbial safety of minimally processed vegetables. *Int. J. Food Sci. Technol.* **34**:1–22.
68. **Francis, G. A., C. Thomas, and D. O'Beirne.** 2002. Effects of vegetable type and antimicrobial dipping on survival and growth of *Listeria innocua* and *E. coli. Int. J. Food Sci. Technol.* **37**:711–718.
69. **Gandhi, M., and K. R. Matthews.** 2003. Efficacy of chlorine and calcinated calcium treatment of alfalfa seeds and sprouts to eliminate *Salmonella. Int. J. Food Microbiol.* **87**:301–306.
70. **Garcia, M. A., M. N. Martino, and N. E. Zarizky.** 1998. Plasticized starch-based coatings to improve strawberry (Fragaria × Ananassa) quality and stability. *J. Agric. Food Chem.* **46**:3758–3767.
71. **Garge, N., J. J. Churey, and D. F. Splittstoesser.** 1990. Effect of processing conditions on the microflora of fresh-cut vegetables. *J. Food Prot.* **53**:701–703.
72. **Golden, D. A., E. J. Rhodehamel, and D. A. Kautter.** 1993. Growth of *Salmonella* spp. in cantaloupe, watermelon, and honeydew melons. *J. Food Prot.* **56**:194–196.
73. **Gomez, R., M. Marina, D. A. Begona, and M. C. Pilar.** 2002. New procedure for the detection of lactic acid bacteria in vegetables producing antibacterial substances. *Lebensm.-Wiss. Technol.* **35**:284–288.
74. **Gonzalez-Aguilar, G., C. Y. Wang, and G. J. Buta.** 2004. UV-C irradiation reduces breakdown and chilling injury of peaches during cold storage. *J. Sci. Food Agric.* **84**:415–422.
75. **Gonzalez-Aguilar, G., C. Y. Wang, J. G. Buta, and D. T. Krizek.** 2004. Use of UV-C irradiation to prevent decay and maintain postharvest quality of ripe "Tommy Atkins" mangos. *Int. J. Food Sci. Technol.* **36**:775–782.
76. **Gorny, J. R.** 1997. A summary of CA and MA recommendations for selected fresh-cut fruits and vegetables. *Proc. Controlled Atmos. Res. Conf.* **5**:30–66.
77. **Gorris, L. G. M., Y. Witte, and E. J. Smid.** 1994. Storage under moderate vacuum to prolong the keepability of fresh vegetables and fruits. *Acta Hortic.* **368**:476–486.
78. **Gulati, B. R., P. B. Allwood, C. W. Hedberg, and S. M. Goyal.** 2005. Efficacy of commonly used disinfectants for the inactivation of calicivirus on strawberry, lettuce, and a food-contact surface. *J. Food Prot.* **64**:1430-1434.
79. **Guo, X., J. Chen, R. E. Brackett, and L. R. Beuchat.** 2001. Survival of *Salmonellae* on and in tomato plants from the time of inoculation at flowering and early stages of fruit development through fruit ripening. *Appl. Environ. Microbiol.* **67**:4760-4764.

80. **Hagenmaier, R. D., and R. A. Baker.** 1997. Low-dose irradiation of cut-iceberg lettuce in modified atmosphere packaging. *J. Agric. Food Chem.* **45**:2864–2868.
81. **Hampson, B. C., and D. Kaiser.** 1995. Air quality in the food-processing environment: a cleanable HEPA filtration system. *Dairy Food Environ. Sanitation* **15**:371–374.
82. **Han, Y., J. D. Floros, R. H. Linton, S. S. Nielsen, and P. E. Nelson.** 2002. Response surface modeling for the inactivation of *Escherichia coli* O157:H7 on green peppers *(Capsicum annuum)* by ozone gas treatment. *Food Microbiol. Safety* **62**:1188–1193.
83. **Han, Y., D. M. Sherman, R. H. Linton, S. S. Nielson, and P. E. Nelson.** 2000. The effects of washing and chlorine dioxide gas on survival and attachment of *Escherichia coli* O157:H7 to green pepper surfaces. *Food Microbiol.* **17**:521–533.
84. **Han, Y., R. H. Linton, S. S. Nielsen, and P. E. Nelson.** 2004. Reduction of *Listeria monocytogenes* on green peppers (*Capsicum annuum* L.) by gaseous and aqueous and chlorine dioxide and water washing and its growth at 7°C. *Int. Assoc. Food Prot.* **64**:1730-1738.
85. **Han, Y., T. L. Selby, K. K. Schultze, P. E. Nelson, and R. H. Linton.** 2004. Decontamination of strawberries using batch and continuous chlorine dioxide gas treatments. *J. Food Prot.* **67**:2450–2455.
86. **Hedlung, K.** 2002. Soil microbial community structure in relation to vegetation management on former agricultural land. *Soil Biol. Biochem.* **34**:1299–1307.
87. **Hillman, P., F. Gebremedhin, and R. Warner.** 1992. Ventilation system to minimize airborne bacteria, dust, humidity and ammonia in calf nurseries. *J. Dairy Sci.* **75**:1305–1312.
88. **Holzapfel, W., H. R. Geisen, and U. Schillinger.** 1995. Biological preservation of foods with reference to protective cultures, bacteriocins and food-grade enzymes. *Int. J. Food Microbiol.* **24**:343–362.
89. **Howard, L. R., and R. W. Buescher.** 1989. Cell wall characteristics of gamma-radiated refrigerated cucumber pickles. *J. Food Sci.* **54**:1266–1268.
90. **Howard, L. R., G. H. Miller, and A. B. Wagner.** 1995. Microbiological, chemical and sensory changes in irradiated Pico de Gallo. *J. Food Sci.* **60**:461–464.
91. **Itoh, Y., Y. Sugita-Konishi, F. Kasuga, M. Iwaki, Y. Hara-Kudo, N. Saito, Y. Noguchi, H. Konuma, and S. Kumagai.** 1998. Enterohemorrhagic *Escherichia coli* O157:H7 present in radish sprouts. *Appl. Environ. Microbiol.* **64**:1532–1535.
92. **Izumi, H.** 1999. Electrolyzed water as a disinfectant for fresh-cut vegetables. *J. Food Sci.* **64**:536–539.
93. **Jacobs, J. L., and G. W. Sundin.** 2001. Effect of solar UV-B radiation on a phyllosphere bacterial community. *Appl. Environ. Microbiol.* **67**:5488–5496.
94. **Jacxsens, L., F. Devlieghere, P. Ragaert, E. Vanneste, and J. Debevere.** 2003. Relation between microbiological quality, metabolite production and sensory quality of equilibrium modified atmosphere packaged fresh-cut produce. *Int. J. Food Microbiol.* **83**:263–280.
95. **Jenner, P. M., E. C. Hagan, J. M. Taylor, E. L. Cook, and O. G. Fitzhugh.** 1964. Food flavourings and compounds of related structure. I. Acute oral toxicity. *Food Cosmetol. Toxicol.* **2**:327–343.
96. **Johannessen, G. S., S. Loncarevic, and H. Kruse.** 2002. Bacteriological analysis of fresh produce in Norway. *Int. J. Food Microbiol.* **77**:199–204.
97. **Kakiomenou, K., C. Tassou, and G. Nychas.** 1996. Microbiological, physiochemical and organoleptic changes of shredded carrots stored under modified storage. *Crit. Rev. Food Sci. Nutr.* **28**:1–30.

98. **Kangasjarvi, J., J. Talvinen, M. Ultriainen, and R. Karjalainen.** 1994. Plant defense systems induced by ozone. *Plant Cell Environ.* **17**:783–794.
99. **Kim, C., Y. C. Hung, and R. E. Brackett.** 2000. Roles of oxidation-reduction potential (ORP) in electrolyzed oxidizing (EO) water and chemically modified water for the inactivation of food-related pathogens. *J. Food Prot.* **63**:19–24.
100. **Kim, G., D. Kim, D. Cho, and S. Cho.** 2003. Bactericidal effect of TiO_2 photocatalyst on selected food-borne pathogenic bacteria. *Chemosphere* **52**:277–281.
101. **Kim, J., A. E. Yousef, and S. Dave.** 1999. Application of ozone and enhancing the microbiological safety and quality of foods: a review. *J. Food Prot.* **62**:1071–1087.
102. **Koseki, S., and K. Itoh.** 2004. Effect of nitrogen gas packaging on the quality and microbial growth of fresh-cut vegetables under low temperatures. *Int. Assoc. Food Prot.* **65**:326–332.
103. **Koseki, S., and K. Itoh.** 2004. Prediction of microbial growth in fresh cut vegetables treated with acidic electrolyzed water during storage under various temperature conditions. *Int. Assoc. Food Prot.* **64**:1935–1942.
104. **Koseki, S., S. Isobe, and K. Itoh.** 2004. Efficacy of acidic electrolyzed water ice for pathogen control on lettuce. *J. Food Prot.* **67**:2544–2549.
105. **Koseki, S., K. Yoshida, S. Isobe, and K. Itoh.** 2005. Efficacy of acidic electrolized water for microbial decontamination of cucumbers and strawberries. *J. Food Prot.* **67**:1247–1251.
106. **Kubota, Y., T. Shuin, C. Kawasaki, M. Hosaka, H. Kitamura, R. Cai, H. Sakai, K. Hashimoto, and A. Fujishima.** 1994. Photokilling of T-24 human bladder cancer cells with titanium dioxide. *Br. J. Cancer* **70**:1107–1111.
107. **Lee, S. Y., M. Costello, and D. H. Kang.** 2004. Efficacy of chlorine dioxide gas as a sanitizer of lettuce leaves. *J. Food Prot.* **67**:1371–1376.
108. **Leistner, L.** 2000. Basic aspects of food preservation by hurdle technology. *Int. J. Food Microbiol.* **55**:181–186.
109. **Leverentz, B., J. J. Wojciech, W. S. Conway, R. A. Saftner, Y. Fuchs, C. E. Sams, and M. J. Camp.** 2000. Combining yeasts or a bacterial biocontrol agent and heat treatment to reduce postharvest decay of "Gala" apples. *Postharvest Biol. Technol.* **21**:87–94.
110. **Leverentz, B., W. S. Conway, M. J. Camp, W. J. Janisiewicz, T. Abuladze, M. Yang, R. Saftner, and A. Sulakvelidze.** 2003. Biocontrol of *Listeria monocytogenes* on fresh-cut produce by treatment with lytic bacteriophages and a bacteriocin. *Appl. Environ. Microbiol.* **69**:4519–4526.
111. **Li, Y., R. E. Brackett, J. Chen, and L. R. Beuchat.** 2001. Survival and growth of *Escherichia coli* O157:H7 inoculated onto cut lettuce before heating in chlorinated water, followed by storage at 5 or 15 degrees C. *J. Food Prot.* **64**:305–309.
112. **Liao, C. H., D. E. McCallus, and J. M. Wells.** 1993. Calcium-dependent pectate lyase production in the softrotting bacterium *Pseudomonas fluorescens*. *Phytopathology* **83**:813–818.
113. **Liao, C. H., and P. H. Cooke.** 2001. Response to trisodium phosphate of *Salmonella* Chester attached to fresh-cut green pepper slices. *Can. J. Microbiol.* **47**:25–32.
114. **Li-Cohen, A. E., and C. M. Bruhn.** 2002. Safety of consumer handling of fresh produce from the time of purchase to the plate: a comprehensive consumer survey. *J. Food Prot.* **65**:1287–1296.

115. **Lin, C.-M., S. S. Moon, M. P. Doyle, and K. H. McWatters.** 2002. Inactivation of *Escherichia coli* O157:H7, *Salmonella enterica* serotype Enteritidis, and *Listeria monocytogenes* on lettuce by hydrogen peroxide and lactic acid and by hydrogen peroxide with mild heat. *J. Food Prot.* **65:**1215–1220.

116. **Liu, H., Y. Du, X. Wang, and L. Sun.** 2004. Chitosan kills bacteria through cell membrane damage. *Int. J. Food Microbiol.* **95:**147–155.

117. **Lukasik, J., M. L. Bradley, T. M. Scott, M. Dea, A. Koo, W. Hsu, J. A. Bartz, and S. R. Farrah.** 2003. Reduction of poliovirus, bacteriophages, *Salmonella montevideo*, and *Escherichia coli* O157:H7 on strawberries by physical and disinfectant washes. *J. Food Prot.* **66:**188–193.

118. **Lunden, J., T. Autio, A. Markkula, S. Hellstrom, and H. Korleala.** 2003. Adaptive and cross-adaptive responses of persistent and non-persistent *Listeria moncytogenes* strains to disinfectants. *Int. J. Food Microbiol.* **82:**265–272.

119. **Magkos, F., F. Arvaniti, and A. Zampelas.** 2003. Putting the safety of organic food into perspective. *Nutr. Res. Rev.* **16:**211–221.

120. **Mcguire, R. G., and R. D. Hagenmaier.** 2004. Shellac formulations to reduce epiphytic survival of coliform bacteria on citrus fruit post harvest. *Int. Assoc. Food Prot.* **66:**1756–1760.

121. **McKellar, R. C., J. Odumeru, T. Zhou, A. Harrison, D. G. Mercer, J. C. Young, X. Lu, J. Boulter, P. Piyasena, and S. Karr.** 2004. Influence of a commercial warm chlorinated water treatment and packaging on the shelf-life of ready-to-use lettuce. *Food Res. Int.* **37:**343–354.

122. **Meneley, J. C., and M. E. Stanghelline.** 1974. Detection of enteric bacteria within locular tissue of healthy cucumbers. *J. Food Sci.* **39:**1267–1268.

123. **Mereghetti, L., R. Quentin, N. Marquet-Van Der Mee, and A. Audurier.** 2000. Low sensitivity of *Listeria monocytogenes* to quaternary ammonium compounds. *Appl. Environ. Microbiol.* **66:**5083–5086.

124. **Michaels, B., C. Keller, M. Blevins, G. Paoli, T. Ruthman, E. Todd, and C. J. Griffith.** 2004. Prevention of food worker transmission of foodborne pathogens: risk assessment and evaluation of effective hygiene intervention strategies. *Food Serv. Technol.* **4:**31–49.

125. **Michaels, B., V. Gangar, H. Schattenberg, M. Blevins, and T. Ayers.** 2003. Effectiveness of cleaning methodologies used for removal of physical, chemical and microbiological residues from produce. *Food Serv. Technol.* **3:**9–15.

126. **Mukherjee, A., D. Speh, E. Dyck, and F. Diez-Gonzales.** 2004. Preharvest evaluation of coliforms, *Escherichia coli*, *Salmonella,* and *Escherichia coli* O157:H7 in organic and conventional produce grown by Minnesota farmers. *J. Food Prot.* **67:**894–900.

127. **Nakamura, S., and A. Hatanaka.** 2002. Green-leaf-derived c6-aroma compounds with potent antibacterial action that act on both gram-negative and gram-positive bacteria. *Agric. Food Chem.* **50:**7639–7644.

128. **Nguyen-the, C., and F. Carlin.** 1994. The microbiology of minimally processed fresh fruits and vegetables. *Crit. Rev. Food Sci. Nutr.* **34:**371–401.

129. **Nortje, G. L., E. Nel, E. Jordan, R. T. Naude, W. H. Holzafel, and R. J. Grimbeek.** 1989. A microbiological survey of fresh meat in the supermarket trade. Part I. Carcasses and contact surfaces. *Meat Sci.* **25:**81–97.

130. **Nortje, G. L., E. Nel, E. Jordan, R. T. Naude, W. H. Holzafel, and R. J. Grimbeek.** 1990. A quantitative survey of a meat production chain to determine the microbial profile of the final product. *J. Food Prot.* **53:**411–417.

131. **Nuanualsuwan, S., T. Mariam, S. Himathongkham, and D. O. Cliver.** 2002. Ultraviolet inactivation of feline calicivirus, human enteric viruses and coliphage. *Photochem. Photobiol.* **76:**406–410.

132. **O'Connor-Shaw, R. E., R. Roberts, A. L. Ford, and S. M. Nottingham.** 1994. Shelf life of minimally processed honeydew, kiwifruit, papaya, pineapple, and cantaloupe. *J. Food Sci.* **59:**1202–1206, 1215.

133. **Odlaug, T. E.** 1981. Antimicrobial activity of halogens. *J. Food Prot.* **44:**608–613.

134. **Odumeru, J. A., J. Boulter, K. Knight, X. Lu, and R. McKellar.** 2003. Assessment of a wash treatment with warm chlorinated water to extend the shelf life of ready-to-use lettuce. *J. Food Quality* **26:**197–210.

135. **Ogden, I. D., D. R. Fenlon, A. J. A. Vinten, and D. Lewis.** 2001. The fate of *Escherichia coli* O157 in soil and its potential to contaminate drinking water. *Int. J. Food Microbiol.* **66:**111–117.

136. **Ollis, D. F., and H. El-Akabi.** 1993. *Photocatalytic Purification and Treatment of Water and Air.* Elsevier Science, Amsterdam, The Netherlands.

137. **Ordonez, J. A., M. A. Aguilera, M. L. Garcia, and B. Sanz.** 1987. Effect of combined ultrasonic and heat treatment (thermoultrasonication) on the survival of a strain of *Staphylococcus aureus. J. Dairy Res.* **54:**61–67.

138. **Orsat, V., Y. Gariepy, G. S. V. Raghaven, and D. Lyew.** 2001. Radio-frequency treatment for ready-to-eat carrots. *Food Res. Int.* **34:**527–536.

139. **Park, C. M., and L. R. Beuchat.** 1999. Evaluation of sanitizers for killing *Escherichia coli* O157:H7, *Salmonella,* and naturally occurring microorganisms on cantaloupes, honeydew melons, and asparagus. *Dairy Food Environ. Sanitation* **19:**842–847.

140. **Park, C. M., Y. C. Hung, M. P. Doyle, G. O. I. Ezeike, and C. Kim.** 2001. Pathogen reduction and quality of lettuce treated with electrolyzed oxidizing and acidified chlorinated water. *Food Microbiol. Safety* **66:**1368–1372.

141. **Pingulkar, K., A. Kamat, and D. Bongirwar.** 2001. Microbiological quality of fresh leafy vegetables, salad components and ready-to-eat salads: an evidence of inhibition of *Listeria monocytogenes* in tomatoes. *Int. J. Food Sci. Nutr.* **52:**15–23.

142. **Poubol, J., and H. Izumi.** 2005. Shelf life and microbial quality of fresh-cut mango cubes stored in high CO2 atmospheres. *J. Food Sci.* **70:**69–74.

143. **Prakash, A., P. Inthajak, H. Huibregtse, F. Caporaso, and D. M. Foley.** 2000. Effects of low-dose gamma irradiation and conventional treatments on shelf life and quality characteristics of diced celery. *J. Food Sci.* **65:**1070–1075.

144. **Prusky, D., D. Eshel, I. Kobiler, N. Yakoby, D. Beno-Moualem, M. Ackerman, Y. Zuthji, and R. B. Arie.** 2001. Postharvest chlorine treatments for the control of the persimmon black spot disease caused by *Alternaria alternata. Postharvest Biol. Technol.* **22:**271–277.

145. **Raiden, R. M., S. S. Sumners, J. D. Eifert, and N. D. Pierson.** 2003. Efficacy of detergents in removing *Salmonella* and *Shigella* spp. from the surface of fresh produce. *J. Food Prot.* **66:**2210–2215.

146. **Rajapakse, N. C., E. W. Hewett, N. H. Banks, and D. J. Cleland.** 1992. Vacuum infiltration with calcium chloride influences oxygen distribution in apple fruit flesh. *Postharvest Biol. Technol.* **1**:221–229.
147. **Rayner, J., R. Veeh, and J. Flood.** 2004. Prevalence of microbial biofilms on selected fresh produce and household surfaces. *Int. J. Food Microbiol.* **95**:29–39.
148. **Reina, L. D., H. P. Fleming, and E. G. Humphries.** 1995. Microbiological control of cucumber hydrocooling water with chlorine dioxide. *J. Food Prot.* **58**:541–546.
149. **Riordan, D. C. R., G. M. Sapers, and B. A. Annous.** 2000. The survival of *Escherichia coli* O157:H7 in the presence of *Penicillium expansum* and *Glomerella cingulata* in wounds on apple surfaces. *J. Food Prot.* **63**:1637–1642.
150. **Robbs, P., J. Bartz, G. McFie, and N. Hodge.** 1996. Potential inoculum sources for decay of fresh cut celery. *J. Food Sci.* **61**:449–452, 455.
151. **Roberts, R. G., and S. T. Reymond.** 1994. Chlorine dioxide for reduction of postharvest pathogen inoculum during handling of tree fruits. *Appl. Environ. Microbiol.* **60**:2864–2868.
152. **Rodgers, S. K., J. N. Cash, M. Siddiq, and E. T. Ryser.** 2004. A comparison of different chemical sanitizers for inactivating *Escherichia coli* O157:H7 and *Listeria monocytogenes* in solution and on apples, strawberries, lettuce and cantaloupe. *J. Food Prot.* **67**:721–731.
153. **Roller, S., and P. Seedhar.** 2002. Carvacrol and cinnamic acid inhibit microbial growth in fresh-cut melon and kiwifruit at 4°C and 8°C. *Lett. Appl. Microbiol.* **35**:390–394.
154. **Rooner, A. B., and A. C. L. Wong.** 1993. Biofilm development and sanitizer inactivation of *Listeria monocytogenes* and *Salmonella typhimurium* on stainless steel and Buna-n rubber. *J. Food Prot.* **56**:750–758.
155. **Rowbury, R. J.** 1995. An assessment of environmental factors influencing acid tolerance and sensitivity in *Escherichia coli*, *Salmonella* spp. and other enterobacteria. *Lett. Appl. Microbiol.* **20**:333–337.
156. **Rushing, J. W.** 2001. A case study of salmonellosis linked to the consumption of fresh market tomatoes and the development of a HACCP program. *HortScience* **36**:29–32.
157. **Samish, Z., R. Etinger-Tulzynsky, and M. Bick.** 1963. The microflora within the tissue of fruits and vegetables. *J. Food Sci.* **28**:259–266.
158. **Sapers, G. M.** 2001. Efficacy of washing and sanitizing methods for disinfection of fresh fruit and vegetable products. *Food Technol. Biotechnol.* **39**:305–311.
159. **Sapers, G. M., L. Garzarella, and V. Pilizota.** 1990. Application of browning inhibitors to cut apple and potato by vacuum and pressure infiltration. *J. Food Sci.* **55**:1049–1053.
160. **Sapers, G. M., R. L. Miller, and A. M. Mattrazzo.** 1999. Effectiveness of sanitizing agents in inactivating *Escherichia coli* in golden delicious apples. *J. Food Sci.* **64**:734–737.
161. **Sapers, G. M., R. L. Miller, M. Jantschke, and A. M. Mattrazzo.** 2000. Factors limiting the efficacy of hydrogen peroxide washes for decontamination of apples containing *Escherichia coli*. *J. Food Sci.* **65**:529–532.
162. **Sapers, G. M., P. N. Walker, J. E. Sites, B. A. Annous, and D. R. Eblen.** 2003. Vapor-phase decontamination of apples inoculated with *Escherichia coli*. *J. Food Sci.* **68**:1003–1007.

163. **Sarig, P., T. Zahavi, Y. Zutkhi, S. Yannai, N. Lisker, and R. Ben-Arie.** 1996. Ozone for control of post-harvest decay of table grapes caused by *Rhizopus stolonifer*. *Physiol. Mol. Plant Pathol.* **48**:403–415.

164. **Savard, T., C. Beaulieu, I. Boucher, and C. P. Champagne.** 2004. Antimicrobial action of hydrolysed chitosan against spoilage yeasts and lactic acid bacteria of fermented vegetables. *Int. Assoc. Food Prot.* **65**:828–833.

165. **Scott, V. N.** 1989. Interaction of factors to control microbial spoilage of refrigerated foods. *J. Food Prot.* **52**:431–435.

166. **Seo, K. H., and J. F. Frank.** 1999. Attachment of *Escherichia coli* O157:H7 into lettuce leaf surface and bacterial viability in response to chlorine treatments. *J. Food Prot.* **62**:3–9.

167. **Seymour, I. J., D. Burfoot, R. L. Smith, L. A. Cox, and A. Lockwood.** 2002. Ultrasound decontamination of minimally processed fruits and vegetables. *Int. J. Food Sci. Technol.* **37**:547–557.

168. **Seymour, I. J., and H. Appleton.** 2001. Foodborne viruses and fresh produce. *J. Appl. Microbiol.* **91**:759–773.

169. **Sharma, M., and L. R. Beuchat.** 2003. Sensitivity of *Escherichia coli* O157:H7 to commercially available alkaline cleaners and subsequent resistance to heat and sanitizers. *Appl. Environ. Microbiol.* **70**:1795–1803.

170. **Sherba, G., R. M. Weigel, and W. D. O'Brien, Jr.** 1991. Quantitative assessment of the germicidal efficacy of ultrasonic energy. *Appl. Environ. Microbiol.* **57**:2079–2084.

171. **Sinde, E., and J. Carballo.** 2000. Attachment of *Salmonella* spp. and *Listeria monocytogenes* to stainless steel, rubber and polytetrafluorethylene: the influence of free energy and the effect of commercial sanitizers. *Food Microbiol.* **17**:439–447.

172. **Singh, P. K., M. R. Parsek, E. P. Greenberg, and M. J. Welsh.** 2002. A component of innate immunity prevents bacterial biofilm development. *Nature* **417**:552–555.

173. **Smid, E. J., L. Hendriks, H. A. M. Boerrigter, and L. G. M. Gorris.** 1996. Surface disinfection of tomatoes using the natural plant compound transcinnamaldehyde. *Postharvest Biol. Technol.* **9**:343–350.

174. **Smilanick, J. L., C. Crisosto, and F. Mlikota.** 1999. Postharvest use of ozone on fresh fruit. *Perishables Handling Q.* **99**:10–14.

175. **Smith, S., M. Dunbar, D. Tucker, and D. W. Schaffner.** 2003. Efficacy of a commercial produce wash on bacterial contamination of lettuce in a food service setting. *J. Food Prot.* **66**:2359–2361.

176. **Solomon, E. B., H. J. Pang, and K. R. Matthews.** 2003. Persistence of *Escherichia coli* O157:H7 on lettuce plants following spray irrigation with contaminated water. *J. Food Prot.* **12**:2198–2202.

177. **Solomon, E. B., S. Yaron, and K. R. Matthews.** 2002. Transmission and internalization of *Escherichia coli* O157:H7 from contaminated manure and irrigation water into lettuce plant tissue. *Appl. Environ. Microbiol.* **68**:397–400.

178. **Somers, E. B., J. L. Schoeni, and A. C. L. Wong.** 1994. Effect of trisodium phosphate on biofilm and planktonic cells of *Campylobacter jejuni, Escherichia coli* O157:H7, *Listeria monocytogenes* and *Salmonella typhimurium*. *Int. J. Food Microbiol.* **22**:269–276.

179. **Soriano, J. M., H. Rico, J. C. Moltó, and J. Mañes.** 2002. Assessment of the microbiological quality and wash treatments of lettuce served in University restaurants. *Int. J. Food Microbiol.* **58**:123–128.

180. **Stevens, V., A. Khan, J. Y. Lu, C. L. Wilson, P. L. Pusey, E. C. K. Igwegbe, K. Kabwe, Y. Mafolo, J. Liu, E. Chalutz, and S. Drobys.** 1997. Integration of ultraviolet (UV-C) light with yeast treatment for control of postharvest storage rots of fruits and vegetables. *Biol. Control* **10**:98–103.

181. **Stevenson, S. M. L., S. R. Cook, S. J. Bach, and T. A. McAllister.** 2004. Effect of water source, dilution, storage, and bacterial and fecal loads on the efficacy of electrolized oxidizing water for the control of *Escherichia coli* O157:H7. *J. Food Prot.* **67**:1377–1383.

182. **St. Georges, S. D., and J. J. Feddes.** 1995. Removal of airborne swine dust by electrostatic precipitation. *Can. Agric. Eng.* **37**:103–107.

183. **Stopforth, J. D., J. S. Ikeda, P. A. Kendall, and J. N. Sofos.** 2004. Survival of acid-adapted or non-adapted *Escherichia coli* O157:H7 in apple wounds and surrounding tissue following chemical treatments and storage. *Int. J. Food Microbiol.* **90**:51–61.

184. **Tanner, R. S.** 1989. Comparative and integrated controls of *Botrytis cinerea* on apple with *Trichoderma harzianum*. *Biol. Control* **1**:59–62.

185. **Taormina, P. J., and L. R. Beuchat.** 2002. Survival of *Listeria monocytogenes* in commercial food-processing equipment cleaning solutions and subsequent sensitivity *J. Appl. Microbiol.* **92**:71–80.

186. **Tauxe, R. V.** 1997. Emerging foodborne disease: an evolving public health challenge. *Emerg. Infect. Dis.* **3**:425–434.

187. **Thomas, L., M. R. Clarkson, and J. Delves-Broughton.** 2000. Nisin, p. 463–524. *In* A. S. Naidu (ed.), *Natural Food Antimicrobial Systems*. CRC Press, Boca Raton, Fla.

188. **Thunberg, R. L., T. T. Tran, R. W. Bennet, R. N. Matthews, and N. Belay.** 2004. Microbial evalution of selected fresh produce obtained at retail markets. *Int. Assoc. Food Prot.* **65**:677–682.

189. **Tschape, H., R. Prager, W. Streckel, A. Fruth, E. Tietze, and G. Bohme.** 1995. Verotoxinogenic *Citrobacter freundii* associated with severe gastroenteritis and cases of haemolytic ureamic syndrome in a nursery school: green butter as the infection source. *Epidemiol. Infect.* **114**:441–450.

190. **Ukuku, D. O., and G. M. Sapers.** 2001. Effect of sanitizer treatments on *Salmonella* Stanley attached to the surface of cantaloupe and cell transfer to fresh-cut tissues during cutting practices. *J. Food Prot.* **64**:1286–1291.

191. **Ukuku, D. O., V. Pilizota, and G. M. Sapers.** 2001. Influence of washing treatment on native microflora and *Escherichia coli* population of inoculated cantaloupes. *J. Food Safety* **21**:31–47.

192. **Venkitanarayanan, K. S., G. O. I. Ezeike, Y. C. Hung, and M. P. Doyle.** 1999. Inactivation of *Escherichia coli* O157:H7 and *Listeria monocytogenes* on plastic kitchen cutting boards by electrolyzed oxidizing water. *J. Food Prot.* **62**:857–860.

193. **Vijayakumar, C., and C. E. Wolf-Hall.** 2003. Evaluation of household sanitizers for reducing levels of *Escherichia coli* on iceberg lettuce. *J. Food Prot.* **65**:1646–1650.

194. **Watkins, J., and K. P. Sleath.** 1981. Isolation and enumeration of *Listeria monocytogenes* from sewage, sewage sludge and river water. *J. Appl. Bacteriol.* **50**:1–9.

195. **Wei, C., W. Y. Lin, Z. Zainal, N. E. Williams, K. Zhu, A. P. Kruzic, R. L. Smith, and K. Rajeshwar.** 1994. Bactericidal activity of TiO2 photocatalyst in aqueous media: toward a solar-assisted water disinfection system. *Environ. Sci. Technol.* **28**:934–938.

196. **Weissinger, W. R., and L. R. Beuchat.** 2000. Comparison of aqueous chemical treatments to eliminate *Salmonella* on alfalfa. *J. Food Prot.* **63**:1475–1482.

197. **Wells, J. M., and J. E. Butterfield.** 1997. *Salmonella* contamination associated with bacterial soft rot of fresh fruits and vegetables in the marketplace. *Plant Dis.* **81**:867–872.

198. **Wright, J. R., S. S. Sumner, C. R. Hackney, M. D. Pierson, and B. W. Zoecklein.** 2000. Reduction of *Escherichia coli* O157:H7 on apples using wash and chemical sanitizer treatments. *Dairy Food Environ. Sanitation* **20**:120–126.

199. **Yam, K. L., P. T. Takhistov, and J. Miltz.** 2005. Intelligent packaging: concepts and applications. *J. Food Sci.* **70**:1–10.

200. **Yaun, B. R., S. S. Summer, J. D. Eifert, and J. E. Marcy.** 2004. Inhibition of pathogens on fresh produce by ultraviolet energy. *Int. J. Food Microbiol.* **90**:1–8.

201. **Zagory, D.** 1999. Effects of post-processing handling and packaging on microbial populations. *Postharvest Biol. Technol.* **15**:313–321.

202. **Zhang, S., and J. M. Farber.** 1996. The effects of various disinfectants against *Listeria monocytogenes* on fresh-cut vegetables. *Food Microbiol.* **13**:311–321.

203. **Zhao, T., M. P. Doyle, and P. Zhao.** 2004. Control of *Listeria monocytogenes* in a biofilm by competitive-exclusion microorganisms. *Appl. Environ. Microbiol.* **70**:3996–4003.

204. **Zhuang, R.-Y., L. R. Beuchat, and F. J. Angulo.** 1995. Fate of *Salmonella montevideo* on and in raw tomatoes as affected by temperature and treatment with chlorine. *Appl. Environ. Microbiol.* **61**:2127–2131.

205. **Zhuang, R. Y., and L. R. Beuchat.** 1996. Effectiveness of trisodium phosphate for killing *Salmonella montevideo* on tomatoes. *Lett. Appl. Microbiol.* **22**:97–100.

206. **Zivanovic, S., S. Chi, and A. F. Draughon.** 2005. Antimicrobial activity of chitosan films enriched with essential oils. *J. Food Sci.* **70**:45–51.

Microbiology of Fresh Produce
Edited by Karl R. Matthews

Microbiological Safety of Fresh-Cut Produce: Where Are We Now?

5

Arvind A. Bhagwat

Ready-to-eat, minimally processed fruits and vegetables are ideally suited for a health-conscious, fast-paced society. Ready-to-eat produce is not subjected to any killing steps such as surface pasteurization or cooking, and producers rely heavily on refrigeration temperatures and modified-atmosphere packaging (MAP) to reduce the microbial load. Minimally processed fresh-cut produce, which is often eaten raw, represents a new challenge to food safety. Many of the foodborne pathogens are evolving, and new outbreak strains are emerging that have adapted to new environmental niches. This chapter focuses on understanding and addressing the food safety needs of the fresh-cut-produce sector. Topics covered here are microbial adaptation to selective pressures exerted by fresh-cut-produce preparation methods, the relationship between acid tolerance and the infective doses of enteric human pathogens, the present status of detection of foodborne pathogens in fresh-cut produce, and the role of (government) action agencies in foodborne outbreaks.

The fresh-cut-produce industry has experienced double-digit growth in sales for the past few years and is poised to continue this trend over the next decade. Consumers have made packaged salads the second-fastest-selling item in U.S. grocery stores, trailing only bottled water. With consumers seeking more prepared foods and ingredients, several fresh-cut-produce preparation facilities are well positioned to meet the consumer needs. Under the new dietary guidelines, the U.S. Department of Agriculture (USDA) urges American consumers to eat more fruits and vegetables—five to nine servings per day. To meet the rising demand of on-the-go consumers, a wide variety

ARVIND A. BHAGWAT, Produce Quality and Safety Laboratory, Henry A. Wallace Beltsville Agricultural Research Center, Agricultural Research Service, U.S. Department of Agriculture, 10300 Baltimore Avenue, Bldg. 002, BARC-W, Beltsville, MD 20705-2350.

of ready-to-eat produce items are available (28). Presently, the fresh-cut-produce industry is estimated to have $12 billion in annual sales in the United States, of which about $5 billion are attributed to cut packaged salad and vegetables. Comparatively, the fresh-cut-fruit market is in its infancy and its retail sales are presently valued at $300 million but are expected to surpass $1 billion by 2008 (104, 105).

Several reasons can be cited for the increased consumption of fresh produce. To promote fresh fruit and vegetable consumption among the schoolchildren in the United States, the Nutrition Title of the 2002 Farm Act provided $6 million for the USDA to award to schools to promote the consumption of fresh produce through the Fruit and Vegetable Pilot Program for the 2002 to 2003 school year. The schools reported that 80% of students were very interested in the pilot program and 71% of schools believed that students' interest had increased during the program period (27). A number of consumer market research reports have predicted that the demand for fresh-cut fruit products will continually increase, with food service establishments and school lunch programs being major customers (5, 65). Several fresh-cut products are available in supermarkets and in food service facilities, and fresh-cut apples have been introduced as a component in school lunch programs. Produce is recognized as an important component of a healthy diet because it is a source of vitamins, minerals, fiber, and antioxidants (21). Consumption of produce can play an important role in weight management as well. Rising income, improved diet, and increasing health knowledge of American consumers are also some of the factors contributing to increased demand. As baby boomers are aging, they are becoming more health conscious and want to eat healthy food (21, 105). Additionally, bad publicity associated with meat products because of the link to pathogens such as mad-cow disease, coupled with an increasingly occupied society with less time allocated for cooking and more meals away from home, is fueling the demand further. Today's health-conscious consumers are always on the look-out for new ways to include healthy food in their diets without spending more time in the kitchen. Many epidemiological studies have shown negative correlations between intake of fruits and vegetables and the incidence of several important diseases, including cancer and atherosclerosis (141, 149, 159). The consumption of attractive and convenient healthy snacks, which are ready to eat and minimally processed, is predicted to increase by double digits until 2020, with most growth occurring in the away-from-home sector (104). While the market may be poised to experience huge growth, this does not mean that fulfilling consumer needs will be easy since delivering high-quality, fresh-cut fruits and vegetables is more difficult than providing consumer products with longer shelf lives, such as canned food or bottled water

(157, 158). From farm to fork, the food industry has to race against time in order to maintain fresh-like attributes of plant tissue that is living and injured and begins to deteriorate upon harvest (24, 158).

FRESH-CUT PRODUCE AND VULNERABILITY TO MICROBIAL SPOILAGE

The properties that make fresh-cut fruits and vegetables unique also pose equally unique challenges for their commercial development. Food safety is one of the most important challenges facing public health officials (Fig. 1). The data indicate substantial declines in the incidence of infections caused by *Campylobacter jejuni*, *Escherichia coli* O157:H7, and human parasites (*Cryptosporidium parvum* and *Cyclospora cayetanensis*). However, increased efforts are needed to reduce the incidence of infections caused by *Salmonella* and *Vibrio* spp. (33). Outbreaks of illness from the consumption of intact produce seem to occur less frequently than those from other foods, such as meat and poultry products (3, 56, 73, 131). Nevertheless, the physical and chemical barrier provided by the epidermis, which prevents the establishment of

Figure 1 Incidence of common foodborne cases in the FoodNet catchment area (33). Data were computed based on the total number of reported cases during the period from 1996 to 1998 (1997 to 1998 for parasitic pathogens) divided by the total number of person-years under surveillance. Incidence rates for *Campylobacter*, *E. coli* O157:H7, *Salmonella*, and *Shigella* infections are expressed as cases per 100,000 persons; those for *Listeria*, *Vibrio*, and parasitic-pathogen infections are expressed as cases per 1,000,000 persons.

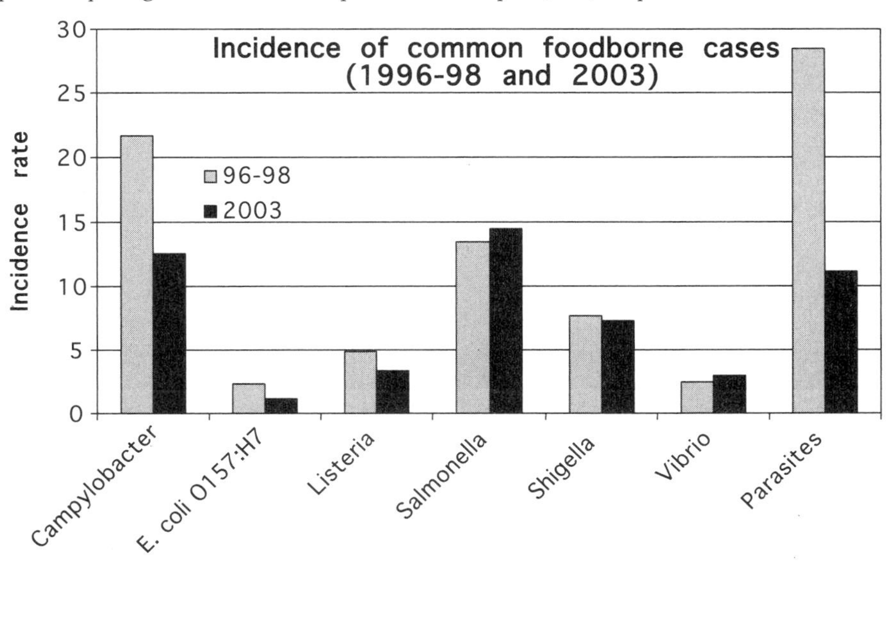

microbes on the surface of produce, is removed during preparation of produce for the fresh-cut-produce market. This may result in the establishment of high populations of foodborne pathogens, leading to increased health hazards. Organic produce has also become increasingly popular because of the belief that fruits and vegetables grown without the application of chemical pesticides and fertilizers are of higher quality and "healthier" than those grown with modern agronomic practices. These claims need scientific validation. Due to the prevalent use of composts and manure in most organic farming, microbial loads on the produce should be examined more closely. Today, the top five fresh-cut-produce processors control 87% of the market share (64). The rapid development of food distribution chains from local manufacturers to national and international markets (28) and new manufacturing methods and modes of preservation (127, 153) may allow the emergence of new pathogens or the expansion of agents previously restricted to particular niches (4, 113). Indeed, the wide distribution of produce from one farm can generate outbreaks of foodborne disease on an international scale (28). In addition to public health concerns of various levels of severity, foodborne outbreaks pose huge economic challenges to manufacturers and consumers (29, 113). Safety regulations, which can differ from one country to another, also have an influence on the international exchange of food products (4, 28).

Fresh-Cut Produce Is Wounded Plant Tissue

Fresh-cut fruits and vegetables are more perishable than their corresponding whole, uncut commodities due to wounding during fresh-cut-produce preparation (22). Preparation of fresh-cut produce involves peeling (carrots; oranges), shredding (carrots; cabbage), cutting (lettuce; apples; pears), and dicing (tomatoes). The impact of these fresh-cut-produce preparation practices is the availability of more plant surface for microbes to attach to and colonize. Microbes attach more easily to the cut or bruised surfaces than to the intact product. Intact cell walls, pectin, and waxing of whole produce prevent microbial contamination. The cut surfaces of fruits and vegetables release significant quantities of liquid that contain nutrients readily utilized by the attached microorganisms. The injured plant tissue and leaking juices also interfere with the sanitizing action of the initial wash steps, making the steps less effective. The exposed plant tissue and readily available nutrients make fresh-cut products vulnerable to cross-contamination at terminal stages in food service establishments (93, 151). Subjecting the plant tissue to fresh-cut-produce preparation practices generates plant physiological responses that resemble responses to wounded tissue. Wounding of plant tissue induces a number of physiological disorders that must be minimized to obtain and maintain fresh quality products.

Wounded Plant Tissue Respires at a Higher Rate

The intensity of the wound response is affected by a great number of factors. The cultivar selection is probably the most important consideration in fresh-cut-produce processing because cultivars can vary greatly in characteristics related not only to texture, flavor, and skin color but also to tissue browning, respiration, and chilling sensitivity. Although all produce commodities respire (breathe), and the rates of respiration differ by commodity, in general fresh-cut produce respires at increased rates compared to whole produce. It is a common practice to keep the concentration of O_2 in a produce package at low levels (1 to 5%) to reduce respiration rates of fruits and vegetables (80). It is generally accepted that reduced O_2 and elevated CO_2 levels generated from MAP can reduce the respiration rate, inhibit browning, retard produce deterioration, and extend shelf life (68, 85). Also, O_2 concentrations below 8% reduce the production of ethylene, a key component of the ripening and maturation process of some produce. MAP is achieved either by constantly monitoring and displacing the gas in the packages or by using passive means, when a desired atmosphere is adjusted at the time of packaging (135).

MAP technology is the key technology widely used in the fresh-cut-produce industry to maintain quality and extend shelf lives of packaged products. But packaging atmospheres with excessively low O_2 and high CO_2 levels can also cause anaerobic respiration, a detriment to product quality and shelf life (66, 109). Anaerobic conditions may also alter the microflora and enhance the growth of anaerobic bacteria (9). To enhance the quality and shelf life, some produce is cut, washed, and stored for several hours at ambient temperature before it is packaged (87). Little is known about how the internal atmosphere of the package impacts the growth of resident microbes, plant pathogens, and human pathogens if present; the interactions of the pathogens with the native microflora; and subsequent food safety and quality. In any case, the main aim of MAP is to extend shelf life, and it must be reiterated that extended product shelf life does not correspond to increased food safety. On the contrary, it gives more time for pathogens (if present) to grow than the short shelf life of fresh-cut produce packaged under ordinary atmospheric conditions (Fig. 2). Lower O_2 concentrations can potentially promote growth of foodborne pathogens such as *Clostridium botulinum* (8) and facultative anaerobic human pathogens (e.g., *E. coli* O157:H7 and *Salmonella* and *Shigella* spp.). To date, two MAP produce products (coleslaw mix and ready-to-eat salad vegetables) have been implicated in outbreaks of botulism and salmonellosis (122, 136).

Fresh-Cut Produce Is Less Permissive of Temperature Abuse

Temperature control is an important factor for maintaining the quality and shelf lives of fresh-cut fruits and vegetables (134). In part because of the

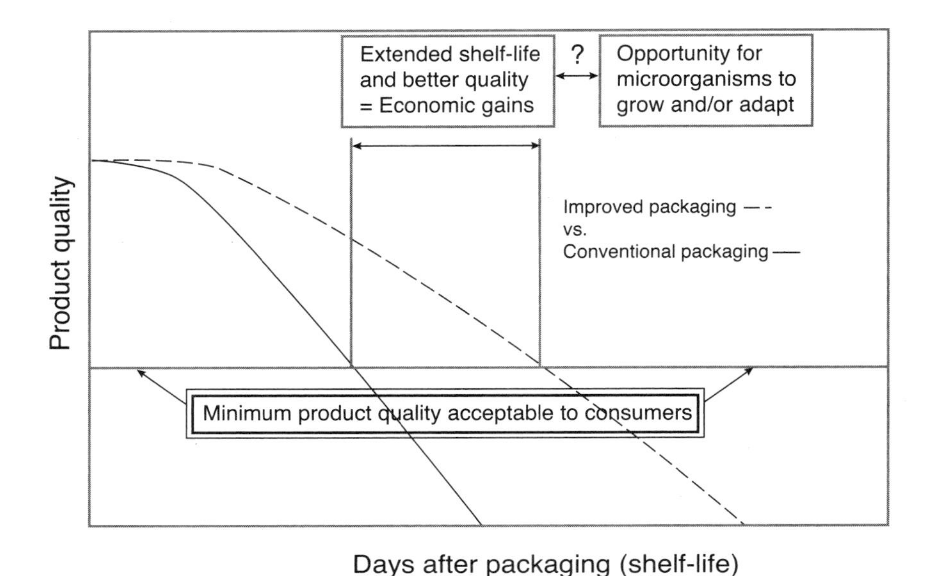

Figure 2 Economic gains versus food safety as a result of improved packaging technology.

following factors, exposed tissue surfaces, highly metabolizing wounded tissue, and easily available nutrients at the wound surface (see above), fresh-cut produce is less permissive of temperature abuse than whole produce. The Food and Drug Administration (FDA) uniform food code requires that all refrigerated hazardous foods be stored at 5°C or below. However, in reality, temperature abuse occurs frequently during fresh-cut-product shipping, transportation, and retail display. Maintaining proper storage temperatures is often most difficult at the retail level, due to the increased handling and the need to display the product in a visually appealing manner. Produce subjected to mechanical injury and abusive temperatures is likely to have high microbial loads when it arrives at the fresh-cut processing facility. Figure 3 illustrates a typical temperature profile as lettuce is harvested and arrives at a retail or fresh-cut processing facility. In a study in which more than 1,400 perishable food items (produce and meat products) were examined for temperature abuse at retail locations, over 87% of samples were being held at temperatures above those recommended (94, 110). Unlike thermally processed foods, fresh-cut products are not sterile, and human pathogens (if present) may coexist with plant pathogens and other indigenous microorganisms. Microbial interactions may play a major role in determining how temperature abuse impacts food quality and microbial safety of fresh-cut products. Presently, only factors associated with food safety risks during pro-

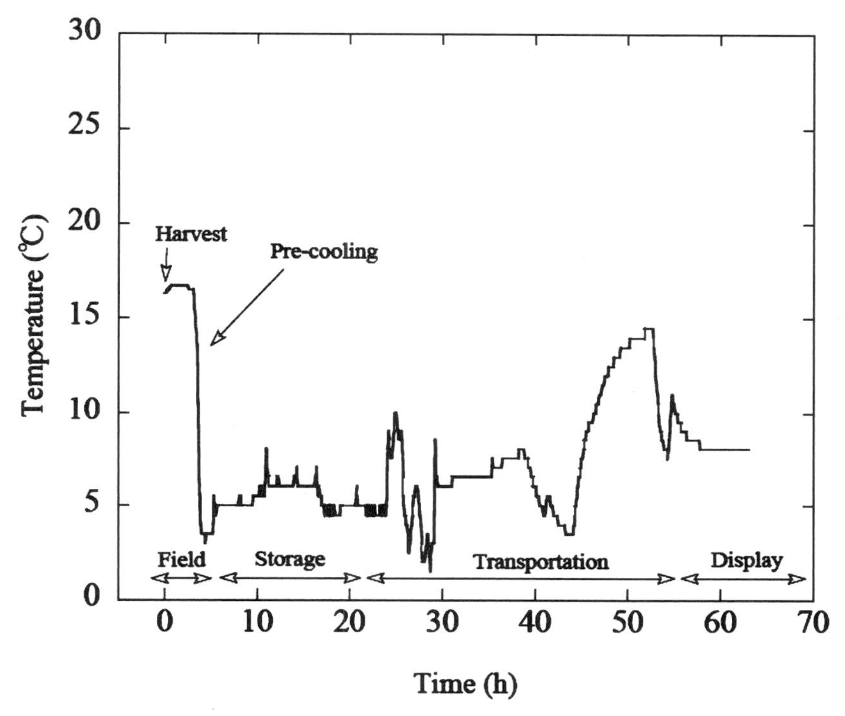

Figure 3 Real-time temperature profile of a typical lettuce shipment from field to consumers. (Data courtesy of S. Koseki et al., National Food Research Institute, Ibaraki, Japan.)

cessing are considered to be "critical control points," or CCP. Therefore, clarification of the impact of temperature on safety and quality will have a fundamental impact on determining appropriate CCP for the hazard analysis of CCP (HACCP) program used by fresh-cut-produce operations.

Many vegetables are sensitive to nonfreezing temperatures (approximately 8 to 10°C and suffer physiological damage if held at chilling temperatures beyond a specific period of time (66). Vegetable commodities include tomato, zucchini, pepper, sweet potato, and jicama. In contrast to those of the whole-produce commodity, shelf life and microbiological safety of fresh-cut produce are severely jeopardized when the produce is stored at temperatures permissive of bacterial growth (above 4°C). Fortunately, development of chilling injury symptoms is less pronounced in ripe tissue and does not occur rapidly. Moreover, most fresh-cut produce is consumed soon after purchase and fruit vegetables (e.g., tomato, eggplant, cucumber) are fully ripe when processed. Since the rate of deterioration resulting from chilling is slower than the rate of microbial spoilage, chilling injury symptoms usually do not develop to a significant extent prior to microbial spoilage of the product. However, the physical damage incurred by the produce during preparation and increased vulnerability to deleterious internal and external changes due to microbial growth require that fresh-cut fruits and vegetables be handled with a greater degree of care than whole intact commodities.

IS FRESH-CUT PRODUCE PREDISPOSED TO MICROBIAL CONTAMINATION?

While meat, fish, and dairy commodities have been typically considered higher risk in terms of food safety, fresh-cut produce is now no longer considered low risk, as evidenced by a number of high-profile foodborne illness outbreaks associated with the consumption of fruits and vegetables (73, 139). Several outbreaks of listeriosis in the United States and Canada have been linked to fresh produce (25, 56, 131). Surveys to determine the prevalence of *Listeria monocytogenes* in various foods indicated that the highest prevalence rate (4.7%) was in seafood salads, and relatively lower prevalence rates were reported for deli salads and bagged salads, at 2.4 and 0.74%, respectively (63). One definition of fresh-cut produce is ready-to-use products which have not been cooked, canned, frozen, or dried for long-term preservation. Pineapple or melon products packed in juice or syrup are considered processed products (not fresh) (146). Thus, microbiological safety is a potential health issue since fresh produce does not undergo a killing step (such as cooking) that would destroy human pathogens that may be present prior to consumption. For the reasons outlined above, fresh-cut-produce preparation practices create unique situations in which microbial contamination may occur. Processing plants may receive their produce from a number of different farms, increasing the risk that different soilborne pathogens may be present. It has been documented that under certain circumstances, many fresh-cut products have served as vehicles for foodborne pathogenic microorganisms and can cause human illness (25, 30, 76). More nonthermal sanitizing practices are being used and developed in order to retain the flavor and texture of the product. Cut produce cannot withstand harsh temperature and sanitizing treatments that have little impact on intact produce. Postharvest microbial spoilage is one of the major causes of quality loss in fresh-cut produce (64).

MICROBIAL QUALITY OF PRODUCE PRIOR TO FRESH-CUT PROCESSING

There is a correlation between initial microbial population, temperature during postharvest processing, and microbiological quality of the finished product (162). However, a low initial population does not guarantee low microbial load during storage. A maximum microbial load of 5 logs per g at the production stage and approximately 7 logs at the use-by date is permitted for fresh-cut produce in Europe (116). In the United States, there is no formal regulation with respect to microbial load for fresh-cut produce, but the U.S. meat industry uses an 8-log load per g as an indicator of spoilage.

SOME PRODUCE HAS UNIQUE CHARACTERISTICS

Cantaloupes and melons have been associated with foodborne outbreaks involving *E. coli* O157:H7, norovirus, and several serovars of *Salmonella* spp. (32, 113). In many of these outbreaks, fruits were cut and had been subjected to temperature abuse. In some cases, melons were contaminated through inadvertent contact with raw meat, packinghouse wash water, or ice used for shipping. In most other instances, contamination was thought to have originated from soil or dirt on the melon rind. Transfer of bacteria from the rind to edible melon flesh can occur (145, 148). *Salmonella* spp. and *E. coli* O157:H7 can grow to high populations on cut melons stored at ambient temperature; however, signs of spoilage are not necessarily evident. Cut cantaloupe is considered to be a potentially hazardous food according to the FDA food code because it is capable of supporting the growth of pathogens due to mild acidity (pH 5.2 to 6.7) and high water activity (0.97 to 0.99). In two independent surveys conducted by the FDA, the frequencies of *Salmonella* isolated from Mexican cantaloupes were 0.78 and 1.08%, with 8 to 12 different serotypes in each survey. More recently, the FDA isolated *Salmonella* from 8 (5.3%) and *Shigella* from 3 (2.0%) of 151 cantaloupe samples collected from nine countries exporting to the United States. Cut melons are subject to temperature and time requirements of the FDA model food code critical for potentially hazardous food. According to a study by Parnell et al. (121), the most efficient method to remove microbes from the surfaces of melons is to scrub the melons with a clean brush under running water. The FDA also recommends that retail establishments wash melons before cutting them and that they clean and sanitize utensils and surfaces of cutting boards when preparing cut melons. Moreover, cut melons should be kept at or below 5°C and should be displayed no longer than 4 h if they are not refrigerated.

PERSONAL HYGIENE DURING FRESH-CUT-PRODUCE PREPARATIONS

It is common for human hands to make contact with fruits and vegetables during harvest and especially during postharvest operations for fresh-cut fruits and vegetables. Hand contact during fresh-cut-produce operations is of particular importance because an infected worker can transfer feces to hands and then to food that is destined to be consumed raw. Farm residents often develop immunity, possibly due to recurrent exposure to less virulent strains of human pathogens, and become asymptomatic carriers of a causative agent (133). For example, of five documented outbreaks associated with orange juice, three have been the result of accidental contamination introduced by an infected person involved in the juice preparation (73). Theoretically, it should not be difficult to control the cleanliness of workers' hands by requiring proper washing or wearing of gloves, but it still remains a difficult task. Another factor that is crucial to the microbiological safety of fresh-cut produce is the sanitization of equipment and containers used for cutting, dicing, and storage. Incidences of cross-contaminations due to common equipment usage for processing of raw meat and preparation of fresh-cut fruits and vegetables have been well documented (93, 151).

ANTIBROWNING TREATMENTS AND FRESH-CUT FRUITS

Preservation of fresh-cut fruits presents unique challenges to the food industry because these products have active metabolisms that can result in rapid tissue deterioration if not controlled (1). Calcium salts, particularly $CaCl_2$, are used as firming agents for a wide variety of whole, peeled, and fresh-cut fruits and vegetables. Calcium propionate, an antimicrobial food preservative (45), was shown to extend the shelf life of fresh-cut apple slices better than other calcium salts tested (26). Reducing agents and acidulants such as ascorbic and isoascorbic acid are widely used to prevent browning of fresh-cut fruits and vegetables (60), including fresh-cut apple slices (129) (Fig. 4). Treatment with high concentrations of calcium salts, reducing agents, and polyphenol oxidase inhibitors prevented browning of fresh-cut apple slices for up to 5 weeks at 5°C in air (26). Browning inhibitor solutions are often expensive, and fresh-cut-produce processors may wish to extend the solution life beyond several batches of produce in a single day or shift. However, during extended periods of use, these solutions may become contaminated with microorganisms, plant tissue, and the juice from previous batches of fruits. Presently, there is no processing method that will totally inactivate pathogens on fresh-cut fruits without altering the quality of the product. The challenge of preventing bacterial growth in processing aids used in fresh-cut-apple preparation practices has recently been addressed (19, 86). The killing of *Listeria innocua* in calcium ascorbate solution by acidification to a low pH with acetic acid was demonstrated, but the treatment had a detrimental effect on apple

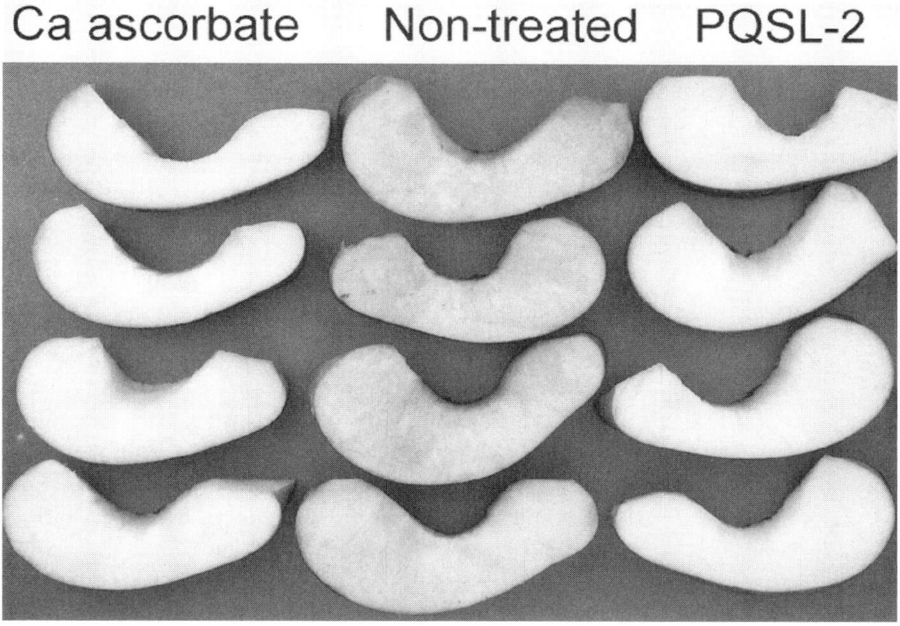

Ca ascorbate Non-treated PQSL-2

Figure 4 Visual quality of fresh-cut apple slices after 7 days of storage at 5°C. Apple slices were treated with either commercial preparations containing Ca ascorbate or Produce Quality & Safety Laboratory—Solution 2 (PQSL-2) wash solution or were left untreated (A. A. Bhagwat and R. A. Saftner, unpublished results; 19).

quality (86). Conversely, an experimental wash solution with activity against five common foodborne pathogens was reported to be comparable with the commercial wash treatment and maintained instrumental and sensory quality characteristics of fresh-cut apples (19). Prior to use with apple slices, the experimental wash solution was shown to reduce the survival of *Salmonella enterica* serovar Typhimurium, *Vibrio cholerae, E. coli* O157:H7, *Listeria monocytogenes,* and *Shigella flexneri* by at least 5 logs, whereas commercially available wash solution had activity only against *V. cholerae*. During treatment of apple slices, the chemical compositions of the wash solutions changed over time as measured by a decrease in conductivity, increases in soluble solid contents and osmolality, and changes in pH. These chemical changes decreased the antibacterial activity of the wash solution. Antibrowning wash solutions should not be reused for multiple batches of sliced apples due to the potential of cross-contamination. Instead, alternative washing strategies that maintain the antimicrobial properties of the wash solutions need to be developed for use on fresh-cut apple slices.

WATER QUALITY AT THE FRESH-CUT-PRODUCE OPERATION FACILITY

Water, either in the form of liquid used to wash the produce or in the form of ice used in shipping or storage, is also a potential source of foodborne microbes. Irrigation water quality is also of significant importance as it may be associated with contamination of produce in the field that is not removed by the time the product reaches the fork (48, 137, 152). *E. coli* O157:H7, *Salmonella*, *Shigella*, *L. monocytogenes*, *Cryptosporidium*, hepatitis A virus, and *Cyclospora* have all been associated with foodborne illnesses traced to poor or unsanitary postharvest practices, specifically to the use of nonpotable cooling water and ice (73, 137). No studies have ever been conducted to determine if the infectivities of enteric pathogens are different for children than for adults. A growing body of evidence, however, indicates that the greater risk of infection with enteric pathogens is for persons less than 19 years of age and is closely associated with the quality of the drinking water these persons consume (118). Washing is an important step during fresh-cut-produce preparation. Washing is the only step aimed at removing foodborne pathogens during the entire fresh-cut-produce preparation process (15, 25). However, washing can present an opportunity for cross-contamination with microorganisms, especially when water is recirculated from different processing lines, a practice that is common in the fresh-cut-produce industry. Although products are often labeled as "washed" or even "triple washed," water quality and the washing process are largely undefined. Chlorine or other types of sanitizers (as discussed in chapter 4) are often added to the wash water, yet they sanitize neither the product nor the wash water effectively. The efficacy of chlorine added to water that is heavily loaded with soil, plant exudates, and microbial contaminants is also questionable. Thus, to improve quality and safety of fresh-cut products, it is important to conduct research to understand how the washing process and water quality impact the microflora of the products and the consequences that these factors have on food quality and safety.

MICROBIAL ADAPTATION TO FRESH-CUT-PRODUCE PREPARATION AND SANITATION PRACTICES

Microorganisms have remarkable adaptive properties enabling them to respond to environmental stresses, such as extreme temperature, high osmolyte (sugar and salt) concentration, and extreme pH. The adaptive mechanisms have led to the development of a number of new and emerging pathogens and the reemergence of organisms that have been problematic in the past. Also, it comes as no surprise that human pathogenic and plant spoilage microorganisms have developed resistance to a number of preservative agents used in the fresh-cut-produce processing industry. In the past two decades, a number of emerging human pathogens have been recognized as causes of foodborne illnesses associated with fresh produce, including parasites (*Cyclospora* and *Cryptosporidium*), viruses (Norwalk-like virus), and bacteria (*E. coli* O157:H7, *L. monocytogenes,* and *Yersinia enterocolitica*).

Adaptive Response—a Challenge To Overcome

Many of the foodborne pathogens are evolving, adapting to new environmental niches. Recently completed sequencing projects with genomes of foodborne pathogens have revealed several sites with genetic polymorphisms even within genes that are involved in core functions of cells, such as survival under stress conditions (107, 108). A case in point is *rpoS*, which encodes the central regulator of general stress resistance (37, 132). Briefly, RpoS is one of the seven different sigma factors in *E. coli*. These sigma factors interact with core components of RNA polymerase to initiate RNA synthesis using promoters that the enzyme recognizes. When nutrients are plentiful and cells are experiencing rapid growth, cellular metabolism and macromolecular synthesis are at their peaks, and the corresponding genes are better recognized by RNA polymerase when it is associated with sigma factor D (σ^D), encoded by *rpoD*. Under conditions of limited growth, such as stationary phase or subjection to an unfavorable environment, RNA polymerase associated with σ^S (encoded by *rpoS*) efficiently translates genes that are necessary to tolerate the stress. In spite of the fact that *rpoS* expression can be induced by many of the conditions used in minimal or light processing of fresh and fresh-cut produce, such as exposure to heat, acid and alkali washes, and sanitizing agents, *rpoS* is found to be commonly mutated. More research is needed to resolve the apparent paradox that mutations are so commonly observed in a gene which is important for bacterial fitness in many environments (Fig. 5). The central role of *rpoS* in adaptation to various environmental stresses has been demonstrated for several enteric human pathogens, including *E. coli* O157:H7, *Salmonella* spp., and *Shigella* spp. (37, 117, 119, 132). In studies in which pathogens such as *E. coli* and *Salmonella* serovar Typhimurium were examined, several sequence variants of *rpoS* were reported and a range of phenotypes attributable to altered *rpoS* function were demonstrated (52, 82, 88, 108). In the most comprehensive single study, 13 out of 58 Shiga toxin-producing *E. coli* strains associated with outbreaks of foodborne illness contained mutations in the *rpoS* gene (155). In a laboratory batch culture study performed using rich media, acquisition of *rpoS* mutations was associated with a growth advantage in stationary phase (117). A possible explanation is that stationary-phase growth advantage-related *rpoS* mutations may help in nutrient scavenging as bacteria enter the stationary phase (52).

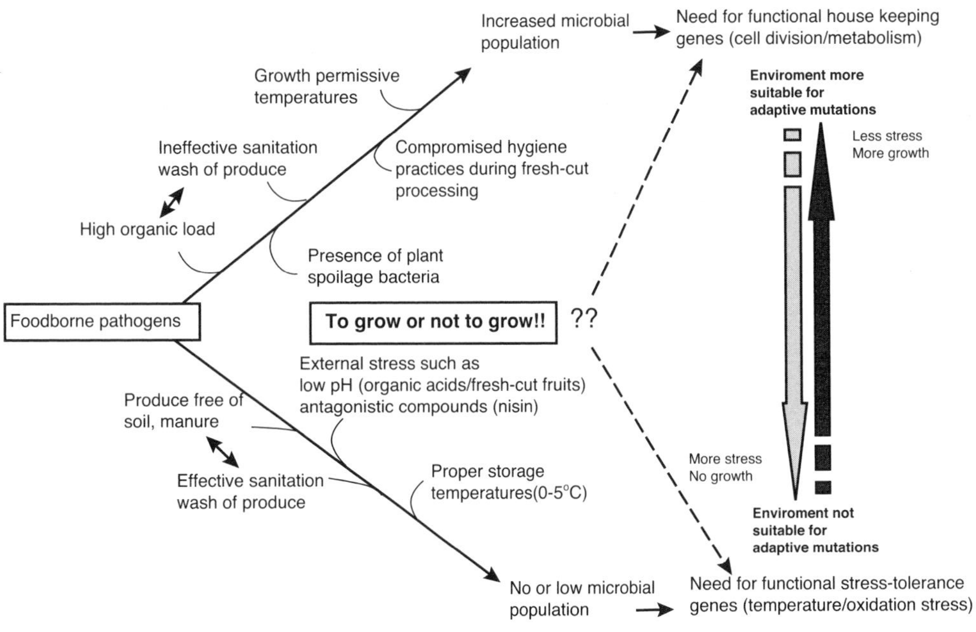

Figure 5 Bacterial dilemma: to grow or not to grow. (Adapted from references 52, 88, and 119.)

Bacterial Dilemma—To Grow or Not To Grow?

In gram-positive microorganisms, the stress-responsive alternative sigma factor, referred to as σ^B, redirects transcription of genes through association with the core RNA polymerase. For example, in *L. monocytogenes*, σ^B contributes to cellular survival under several adverse conditions, such as carbon depletion and exposure to acid, oxidative, low-temperature, and osmolarity stresses. Further investigation on the selective advantages, if any, of loss or mutation of *rpoS* or σ^B and the contribution of these factors to the pathogens' stress tolerance may provide insights into microbial adaptations to various environmental stresses (Fig. 5). It is feasible that microbiological flora residing on fresh and fresh-cut produce, throughout the produce's journey from farm to fork, would undergo cycles of subjection to unfavorable and hostile environments as well as periods of limited growth when produce is stored at abusive temperatures. Similarly, exposure to conditions of improper sanitation may induce stress responses even though the treatments are intended to reduce the microbial load. It may be noted that the deleterious genomic mutation rate in stationary-phase *E. coli* cells is exceedingly high compared with that in fast-growing cells (107). It may be argued that such a high mutation rate may be important for accelerated adaptation and needs to be examined if it represents a response of human pathogens as they survive on fresh and fresh-cut produce under various conditions.

Role of Attachment

Microorganisms in their natural habitat and those present on fresh produce are firmly attached to the substratum and are very likely to be part of a biofilm (84, 123). Examination of common household kitchen items and various produce items revealed widespread existence of microorganisms in biofilms (123). One of the characteristics of bacterial attachment to fruits and vegetables is the rapidity of attachment to the commodity surface. Previous studies suggest that effectiveness of washing is inversely related to the time interval between contamination and washing. In spite of the use of a combination of treatments such as hot water, hydrogen peroxide, and several generally recognized as safe (GRAS) substances, cantaloupes artificially inoculated with nonpathogenic *E. coli* or *Salmonella* strains could not be fully decontaminated after the interval of 1 day between inoculation and wash treatments (128, 144, 145). The reduced efficacy of sanitizer treatments may be due to the strong attachment of and biofilm formation by bacteria on melon surfaces but also may be attributed to the entry of bacteria into sites inaccessible to aqueous sanitizers (144). It was demonstrated previously that *Salmonella* serovar Chester attaches preferentially to the cut surfaces of apples and green peppers, and survives to a much greater extent than other bacteria on the intact apple and pepper surfaces (100, 101). More information is needed on the influence of bacterial attachment to surfaces on infective dose and the way to validate sanitation activity against microorganisms present in biofilms. A more in-depth discussion of foodborne pathogens' attachment to and internalization by fresh fruits and vegetables is provided in chapter 3.

Infective Dose, Attachment, and Acid Tolerance

Information related to several aspects of foodborne pathogens, such as estimated numbers of cases of infection reported annually, infective doses, and the food(s) associated with the illness and food recalls, is made available publicly. Information on the infective dose (the estimated dose necessary to cause sickness or disease) provides a useful reference for the virulent nature of a pathogen in an outbreak and possible insight into the challenges posed by the pathogen in food processing. The infective dose measurement studies are generally based on data obtained from volunteer human feeding studies in which healthy, young adults are used as subjects. As a result, there are sometimes large discrepancies between the infective dose of a given foodborne pathogen as determined from laboratory volunteer studies and that as estimated from epidemiological or outbreak cases (2). The numbers of bacteria consumed in contaminated food can vary considerably, as do the infective doses of enteric pathogens (13, 20, 36, 106). In spite of the diverse virulence characteristics of different enteric human pathogens, one common trait that has emerged very strongly is the ability of these strains to withstand gastric acidity. The low infective dose associated with diarrheagenic *E. coli* strains is attributed to the organisms' acid-resistant nature (Fig. 6). The importance of acid survival in pathogenesis is underscored by the fact that *V. cholerae*, non-Typhi *Salmonella* sp., and *Shigella flexneri* have oral infective doses of 10^9, 10^5, and 10^2 bacteria, respectively (2, 3). These infective doses correlate with the level of acid resistance demonstrated by each organism, with *V. cholerae* being the most acid sensitive and *Shigella flexneri* being the most acid resistant (2-6). Ability to survive the low pH of the human stomach is considered to be an important virulence determinant for *L. monocytogenes*. Moreover, it plays an important role in the survival of this pathogen in a variety of foods.

Salmonella strains have also been shown to develop attachment-mediated acid tolerance (Fig. 7). *Salmonella* spp., which cause primarily foodborne infections, are much less acid resistant than *E. coli* O157:H7 and *Shigella flexneri*, and their infective doses are still a controversial issue (20). In clinical trials in which defined inocula were fed to human volunteers, the infective dose was estimated to be at least 10^6 bacteria (111). At the same time, there are reports from various outbreaks that when the cells were ingested with a food source, *Salmonella* species caused infection at a much lower

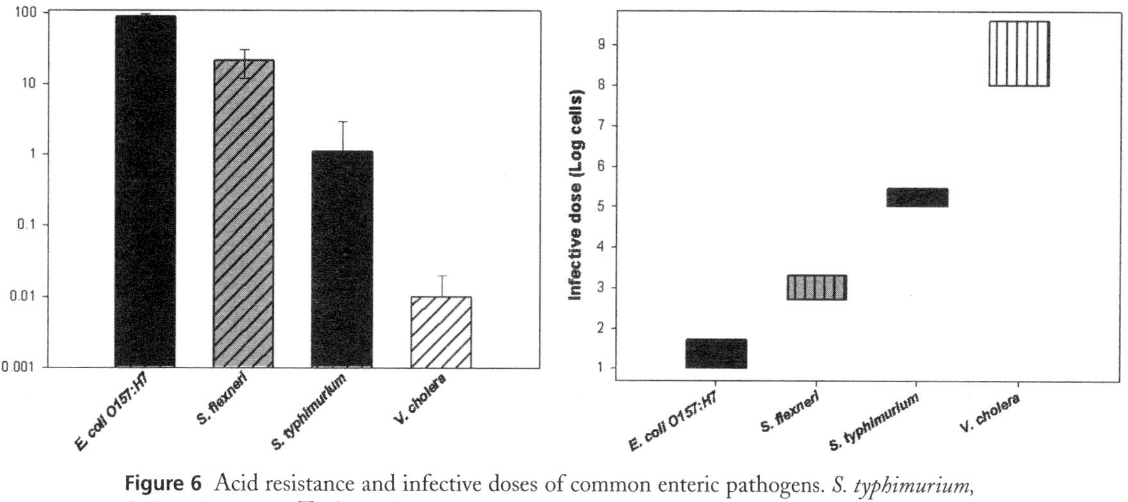

Figure 6 Acid resistance and infective doses of common enteric pathogens. *S. typhimurium*, *Salmonella* serovar Typhimurium.

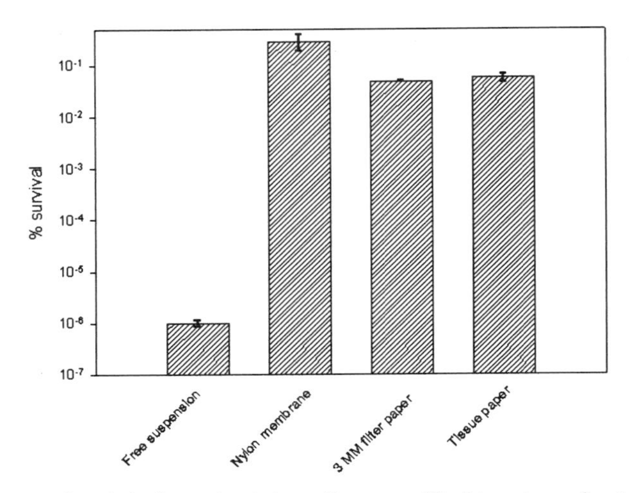

Figure 7 Survival of acid challenge by *Salmonella* serovar Typhimurium after inoculation onto various solid supports (61). Error bars indicate standard deviation.

infective dose (≤100 cells) (41, 42, 74). It has been suggested that solid foods, especially ones rich in fat content, may protect salmonellae against the stomach acidity (41, 154). The biochemical and molecular events leading to the surface-associated protection are not yet fully understood (61, 154). It has been demonstrated that various *Salmonella* strains that would normally be killed during acid challenge (pH 3.0; 37°C; 2 h) survived an identical acid challenge when inoculated onto fresh-cut produce surfaces. The model system developed using cellulose-acetate and nylon membranes would enable researchers to design molecular biology experiments and study the phenomenon of the surface contact-mediated acid protection (61).

Acid tolerance by enteric human pathogens is significant for another reason. Some of the fresh-cut fruits are acidic, and wounded tissue in addition to providing the surface area for attachment and colonization may influence pathogens' ability to withstand gastric acidity and thereby potentially decrease their infective doses. *E. coli* has developed at least three genetically and physiologically distinct acid resistance systems which provide different levels of protection (7, 106). The glutamate-dependent acid resistance (GDAR) system utilizes extracellular glutamate to protect cells during extreme acid challenges, is believed to provide the highest protection from stomach acidity, and can be induced in an *rpoS*-independent manner (7, 31). Many fruits and vegetables have considerable amounts of glutamic acid (Table 1). The GDAR system may also be very important for food safety for yet another reason, as monosodium glutamate in various foods will also support acid resistance, although it is intended to be a food preservative (17a).

Table 1 Glutamic acid contents of various fresh-cut fruits and vegetables[a]

Type of produce	Amt of glutamic acid (mg/100 g of edible portion)
Apple	25
Broccoli	375
Baby carrot	165
Carrot	224
Cauliflower	164
Cucumber with peel	196
Honeydew melon	153
Pineapple	79
Tomato	313
Watermelon	63

[a]Data obtained from reference 147.

MAP and Adaptive Mutations

In order to extend shelf life, more and more ready-to-eat food is marketed in modified-atmosphere packages (135, 160). It is necessary to determine if such an environment would provide semiaerobic conditions for the induction of GDAR in an *rpoS*-independent manner. As stated previously, few studies have directly compared the three acid resistance systems of pathogenic *E. coli*. For example, 13 of 58 toxin-producing *E. coli* isolates from various sources contained mutated *rpoS* genes and exhibited increased acid sensitivity. Recent studies have indicated that some foodborne isolates of diarrheagenic *E. coli* have a mutation(s) in a gene(s) with a yet-undefined role in acid resistance. These results show that in addition to mutant alleles of *rpoS*, mutations in *gadE* exist in natural populations of pathogenic *E. coli* (17a). Such mutations most likely alter infectivity of individual isolates and may play a significant role in determining the infective dose of diarrheagenic *E. coli*.

METHODS TO CONTROL PATHOGENS ON FRESH-CUT PRODUCE

A program that has emerged as a powerful tool to reduce the incidence of foodborne illness is the HACCP system. This approach is proactive rather than reactive, focusing on the evaluation and control of foodborne hazards at every link in the supply chain—from farm to fork. The use of the HACCP program is a regulatory requirement for the seafood, meat and poultry, and juice industries. According to the International Fresh-cut Produce Association (IFPA), it is recommended and prudent for fresh-cut-produce producers to have an HACCP plan in place (64). Accordingly, the HACCP system is commonly used in the fresh-cut-produce industry and many customers require their suppliers to have such a program in place. With respect to foodborne pathogens, such as *E. coli* O157:H7, *Salmonella,* and *L. monocytogenes*, recent food regulatory agency reports have credited the implementation of HACCP programs in regulated industries as a key factor in the reduction in the incidence of microbial contamination of food. *L. monocytogenes*, for example, is a bacterium widespread in the environment, can grow in cold environments such as home refrigerators and food-processing plants, and can cause a deadly disease (listeriosis) in at-risk populations that include pregnant women, immunocompromised individuals, and the elderly (13, 36, 67). In October 2003, the USDA Food Safety and Inspection Service (FSIS) released data showing a 1-year 25% decrease in the occurrence of *L. monocytogenes* in food samples and a 70% decline compared with occurrences in years prior to the implementation of the HACCP program in the meat and poultry industries.

Developing a HACCP plan requires broad knowledge of the food and the food's intended use and hazardous constituents. HACCP plan development also requires detailed understanding of the processes used in the production, packaging, storage, and distribution of the food. The terms HACCP program and food safety program are often used synonymously; however, a HACCP plan is merely a component of an overall food safety program.

Quality of Produce Arriving at the Fresh-Cut-Produce Production Facility

Cleaning and sanitizing are key interventions in the preparation of fresh-cut produce to avoid human pathogen contamination. Conventionally, chlorine has been used as an antimicrobial agent and is normally added to process water used in dump tanks, flumes, and washers to reduce bacterial loads and levels of human pathogens that might be present. This is an effective means to kill planktonic (unattached) cells and to minimize attachment to uncontaminated produce of human pathogens from the process water passing through dump tanks or flumes. Chlorine treatments generally result in 90 to 99% reduction in the microbial load. Although the chlorine washes result in significant increase in the product quality and shelf life, the decrease in the bacterial load is not equivalent to that from surface pasteurization and thus may not be adequate to ensure microbial safety of the ready-to-eat produce. For several decades, chlorine has been the choice sanitation agent and is the agent most widely used to date. Chemicals that are chlorine based are often used to sanitize produce and surfaces within produce-processing and packaging facilities. Chlorine is also used to reduce microbial populations in water used to wash produce and remains a convenient and affordable sanitizer for use against microbial spoilage. However, fresh-cut-produce preparation practices pose a unique challenge. Since chlorine reacts with organic matter, when exposed plant tissue and exudates leak out of plant tissue during dicing, cutting, and peeling, most of the available chlorine is neutralized. As a result, the concentration of available chlorine drops rapidly to sublethal levels. Moreover, cracks and crevices in the produce, along with hydrophobic fruit cuticles, may keep the residing microflora out of reach of the sanitizing agent. The FDA and the National Advisory Committee on Microbiological Criteria for Foods have established a goal of 5-log reduction in the human pathogen population in fresh apple cider and alfalfa sprouts. Although the 5-log-reduction target might not be directly applicable to or required of minimally processed fruits and vegetables, it does indicate the kind of population reduction goal that food safety authorities visualize for ready-to-eat products.

Fruits and vegetables are often immersed in nonpotable water prior to entering the fresh-cut-produce facility. Chlorine, which is normally added at up to 200 ppm (the recommended maximum concentration in a fresh-cut-produce flume system), would be rapidly sequestered in the presence of organic matter which may come from soil, leaves, or farm debris present in the wash water. In fresh-cut-produce preparations, chlorine would rapidly react with components leaking from tissues of cut produce surfaces and the reaction would neutralize some of the chlorine before it reached microbial cells. There are several reports in the literature indicating the inability of chlorine to destroy *Giardia* spp. cysts (14) or to reduce the microbial load by more than 1 to 2 logs for various combinations of pathogens and produce. On fresh-cut cantaloupe cubes, chlorine levels as high as 2,000 ppm could reduce the population of several *Salmonella* spp. by only approximately 1 log (15). A smaller reduction in the viable counts was observed when efficacy was tested on asparagus spears, thereby indicating that it may be necessary to customize sanitizing treatments for different types of produce. The use of chlorine is only a risk reduction factor, and other preventive measures are needed to further reduce the risk of human pathogens on fresh-cut produce.

Limitations of the Use of Chlorine Dioxide

Chlorine dioxide (ClO_2) is a strong oxidizing agent that has about 3.5-fold more oxidative capacity than chlorine. Its high biocidal activity has been confirmed under a wide range of food preparation settings. Moreover, chlorine dioxide does not react with nitrogen-containing compounds or ammonia to form dangerous chloramine compounds as chlorine does. The use of chlorine dioxide gas as a disinfectant or sanitizer has developed during the past 20 years, and the effectiveness of the gas has been validated for medical uses. The effectiveness of chlorine dioxide has been demonstrated against otherwise difficult-to-kill life forms such as spores of *Bacillus* spp. and *Clostridium* spp. (53). In 1998, the FDA approved aqueous ClO_2 for use in washing fruits and vegetables. A maximum of 200 ppm of ClO_2 is allowed for sanitizing of processing equipment, and only 3 ppm maximum is allowable for contact with whole produce. More than a 6-log reduction in levels of the spoilage microorganisms on juice tank surfaces was reported by Han et al. for conditions with less than 90% relative humidity (70). Several researchers have also studied the lethal activity of aqueous ClO_2 on fresh and fresh-cut produce. For example, one study reported a 5-log reduction in levels of *E. coli* ATCC 11229 suspended in water with 1 to 4 ppm of aqueous ClO_2 within 30 s, whereas others have achieved only a 1.5-log reduction (55). These types of data are typically explained by an attachment-mediated increase in stress tolerance. The increased tolerance may be an adaptive response or may be due to inaccessible hydrophobic pockets in the leaves or folds on the produce surface where the aqueous ClO_2 cannot penetrate. However, the use of ClO_2 gas as a sanitizer can overcome the limitations inherent in aqueous sanitizers (70, 71). Researchers have observed an approximately 5-log reduction in the counts of viable *E. coli* O157:H7 cells attached to fresh-cut green peppers following treatment with 1.24 mg of ClO_2/liter for 30 min under 90 to 95% relative humidity (70, 71).

However, there are still some difficulties in large-scale applications of gaseous ClO_2 by the fresh-cut-produce industry. Gaseous ClO_2 must be generated at the site of use, which would require purchasing and operating of complex machinery and additional technical expertise. For the fresh-cut-produce industry, an ideal new sanitizing method needs to be simple and inexpensive. Recently, small chemical sachets were introduced which generate the ClO_2 gas in a simple and convenient way. Lee et al. (95) examined this device and performed experiments with lettuce leaves artificially inoculated with *L. monocytogenes* and *Salmonella* serovar Typhimurium. Gaseous ClO_2 generated from sachets was effective at reducing pathogens on lettuce in the range of 4 to 5 log units without deteriorating the visual quality of the lettuce. More research is needed to standardize this method with other produce commodities and particularly to examine issues such as the effect of humidity and the application of the method to fresh-cut fruits (as there is a potential conflict with antibrowning agents).

Non-Chlorine-Based Sanitizers

Sapers et al. and Ukuku et al. (128, 130, 145) explored the feasibility of hydrogen peroxide treatment for fresh-cut produce, and it appears to be a promising alternative to chlorine. The hydrogen peroxide treatment was recommended on the basis that the shelf life of fresh-cut melons could be extended by 4 to 5 days compared to that of chlorine-treated melons. When hydrogen peroxide was applied at elevated temperatures (50 to 60°C), a reduction of 4 logs in microbial loads was observed. For fresh-cut vegetables, elevated-temperature treatments may provide the additional advantage of inhibiting enzymatic browning occurring at the cut surfaces. Phenylalanine ammonia-lyase is the first enzyme involved in the browning reaction, and its activity is greatly retarded by mild heat treatment applied either before or after cutting. Thus, heat-treated lettuce leaves retain better visual quality than untreated leaves. Similarly, the treatment involving a combination of hydrogen peroxide and mild heat improved the visual quality of fresh-cut cantaloupe (143). The visual quality of cantaloupe cubes prepared from whole fruits treated with 5% hydrogen peroxide at 50°C for 1 min was better than that of cubes prepared from fruits treated with 1,000 ppm of chlorine at 20°C for 1 min (130, 145). Although treatments involving hydrogen peroxide and mild heat (followed by a cold-water rinse) have been shown to lower microbial populations and improve sensory quality, commercial implementation of treatments involving hydrogen peroxide still requires prior FDA approval. Treatments involving a combination of hydrogen peroxide and organic acids such as peroxyacetic acid and lactic acid as antibacterial agents for fresh-cut vegetable surfaces have been suggested (77). Although excellent bactericidal activity against three common pathogens (*Salmonella enterica* serotype Enteritidis, *E. coli* O157:H7, and *L. monocyotgenes*) was observed with such treatment, the sensory quality of lettuce was compromised (99), though that of apples, oranges, and tomatoes remained satisfactory (112, 150). These findings underscore the need for different antimicrobial treatments for different types of produce.

Use of GRAS Substances

Substances that have GRAS status that are added to foods as acidulants and flavor enhancers, for example, sodium and potassium salts of lactic acid, may also have antimicrobial activity. The antimicrobial activity of lactic acid is not solely due to lowering of pH but is also due to the disruption it causes to the outer membranes of gram-negative bacteria. Inclusion of lactic acid at up to 4.8% of the total formulation in fully cooked meats and poultry products is allowed. The application of lactic acid to whole-head lettuce reduced pathogen populations but was detrimental to sensory quality, and consumer response was not favorable (112, 150). Sorbic acid is widely used in foods to inhibit yeast and molds but does not have antibacterial activity. The use of EDTA salts in food is very common, and they are a good color enhancer and inhibit discoloration of food. They are metal ion chelators and exert antibacterial activity by chelating essential divalent cations from bacterial cells.

Nisin, a pore-forming bacteriosin, is active against many gram-positive bacteria. Nisin is more active at the lower pH typical of many fruits and some vegetables and is the only commercially available bacteriocin that is GRAS (140). Its use at different levels has been approved in more than 70 countries for application to a number of food preparations with added fruit and vegetables. Ukuku et al. (144, 145) examined the effects of several GRAS substances and observed significant reduction in the survival of *Salmonella* spp. on fresh-cut melons. However, none of the treatments evaluated totally eliminated the pathogens from the cantaloupe surface or prevented the transfer of *Salmonella* from the whole cantaloupe to the interior flesh when fresh-cut pieces were prepared.

UV and Ionizing Radiation as Antimicrobial Treatments

Consumer acceptance of irradiated food remains questionable. The percentage of consumers that were willing to buy irradiated food declined from 70% in 1996 to 50% in 2000 (59). Produce treated at doses above the level of 1 kGy cannot be labeled as "fresh" according to the FDA (39). Low-dose ionizing radiation treatments have been shown to be effective at achieving 4- to 5-log reduction in counts of viable cells of *L. monocytogenes* on cabbage, tomatoes, and broccoli (11). There was no significant change in the appearance, color, texture, taste, and overall acceptability during 7 days of postirradiation storage at 4°C. Fan et al. (49) combined warm-water treatment with low-dose irradiation (0.5 to 1.0 kGy) to obtain 14 days of shelf life for fresh-cut lettuce.

Irradiation with UV radiation presents an alternative nonthermal processing technique to apple cider manufacturers as well as customers who believe that off flavors are generated during pasteurization of cider. Due to an increase in outbreaks linked to unpasteurized juices and ciders, the FDA initiated new regulations requiring that manufacturers either obtain a minimum 5-log reduction in levels of the pertinent pathogen in the finished product or provide a warning label on the bottle. A recent study has demonstrated that doses of $\geq$6,500 μW s/cm^2 were sufficient to achieve up to 6-log reduction of *E. coli* levels in apple ciders from different cultivars (12).

Combination of Sanitation Agents

Hurdle technology is the deliberate use of multiple preservation techniques in order to establish a series of microbial controls that any microorganisms present should not be able to overcome (96). Researchers have used combinations of chlorine, hydrogen peroxide, electrolyzed water, and various combinations of GRAS substances in designing hurdle treatments. The hurdle concept can then be further extended by the use of modified-atmosphere storage at low temperatures. Koseki and Itoh (91) examined the combination of acidic electrolyzed water treatment, packaging under N_2 gas, and storage at 1°C to effectively extend the shelf life while maintaining produce quality and reducing microbial growth during storage. First, acidic electrolyzed water was used for decontamination of fresh-cut cabbage and lettuce. Second, decontaminated fresh-cut vegetables were enclosed in polyethylene pouches with 100% N_2 atmospheres. Finally, packed fresh-cut vegetables were stored at 1°C, and microbial growth (aerobic bacteria, *Bacillus cereus*, and coliforms) was inhibited for 5 days.

Biocontrol during Preharvest and Postharvest Processing of Produce

Most fresh-cut-fruit and -vegetable cultivars have been developed with little or no effort toward meeting specific fresh-cut-produce requirements. In addition to general cleanliness and hygiene in the field, crop production practices significantly influence fresh-cut-produce quality. Kumar et al. (92) demonstrated that the quality of slices of red tomato fruit from plants grown using hairy vetch mulch was much better than that of slices from plants grown using black polyethylene. The plants grown using organic (hairy vetch) mulch tended to have better disease resistance, and tomato slices had lower levels of electrolyte leakage and eightfold-lower levels of chilling injury (measured as percentage of water-soaking areas). The application of biocontrol agents to prevent proliferation of postharvest spoilage organisms has been studied to a greater extent than control of human pathogens on produce surfaces. Studies suggest that nonpathogenic microorganisms applied to produce surfaces might outcompete pathogens for physical space and nutrients. Additionally, biocontrol agents may produce antagonistic compounds that negatively affect viability of pathogens. There are few published reports on the use of biocontrol agents to prevent growth of human pathogens on fresh-cut fruits and vegetables (72, 125). Bacteriophages possess attributes to control foodborne pathogens in a novel manner by infecting bacterial cells, lysing the cells, and liberating more phage, which in turn can infect the surrounding target bacterium population. Bacteriophages have a history of safe use; they can be highly host specific and replicate in the presence of a host. On the other hand, application of phages as biocontrol agents could be complicated due to an apparent requirement for a threshold level of the host before replication can proceed (81). The use of bacteriophage to reduce populations of *Salmonella* spp. and *L. monocytogenes* on fresh-cut melons and apples was recently reported (97, 98). The *Salmonella* spp. population was not reduced on apple slices due to the fruit's low pH (~4.2); however, better control of *L. monocytogenes* on fresh-cut cantaloupe cubes was observed when phages were applied in combination with nisin. Use of phage for pathogen control deserves further investigation.

Epidemiological Limitations to Trace Back of Contamination

When a foodborne outbreak is suspected, confirmation of microbial contamination, the location of its source, and a list of production batches to be recalled are important information that must be obtained quickly and efficiently. The faster the source of a pathogen can be identified, the sooner the public can regain confidence in the food supply. It is imperative that the food industry maintain consumers' confidence in the food they eat. Most of our knowledge of routes of foodborne transmission of pathogens has been acquired through the study of epidemiological data from various prevalence studies and outbreak investigations (47, 50, 161). Therefore, determining the precise source of contamination is crucial when devising strategies to reduce future outbreaks. However, only 2 of the 27 outbreak investigations on fresh produce clearly identified a point of contamination, which underscores the importance of and need for rapid and accurate pathogen identification methods (115). Several studies have indicated limitations on estimating the causative agent of foodborne illnesses. In one recent study that analyzed cases of foodborne illnesses in 1998 and 1999 that had been catalogued by the FoodNet surveillance system of the Centers for Disease Control and Prevention (CDC), no confirmed etiology was found for 71% of outbreaks and in 45% of cases the suspected food vehicle could not be identified (83). It is estimated that unknown pathogenic agents transmitted in food cause 3,400 deaths per year in the United States (58). Without adequate resources and techniques to study epidemiology, the factors that contribute to these outbreaks will remain poorly understood.

Comparison of Conventional and PCR-Based Detection Methods

Most microbiological methods that are based on enrichment, biochemical characteristics, and immunological identification require 5 to 7 days to complete. Although many rapid-test kits are commercially available, pathogen enrichment is necessary for rapid microbiological, immunological, and real-time DNA-based methods; no direct (i.e., nonenrichment) method has been validated. A comprehensive comparative account of different commercially available detection methods (conventional as well as PCR based) specifically for sprouts is provided in chapter 6.

New Trends in PCR-Based Detection Methods

Advances in biotechnology have permitted more reliable microbial identification and surveillance (51), and PCR-based detection methods have become valuable tools for investigating foodborne outbreaks and identifying the responsible etiological agents (78, 79). Several commercially available PCR-based systems utilize the amplification of a specific target DNA sequence for the detection of various foodborne pathogens (10). The reliability of PCR detection methods depends, in part, on the presence of sufficient numbers of target molecules (51). Because foods comprise complex matrices, most commercially available detection systems require selective enrichment steps to overcome problems of low pathogen populations present in foods (78). Since enteric pathogens are of major concern and can be found as contaminants in similar types of food, it was reasoned that one universal identification scheme would make the task of food surveillance much more manageable. Presently, as per the AOAC- and the FDA-approved protocols, to detect each pathogen, it is necessary to perform the enrichment steps best suited for the individual pathogen (23). To address this issue, a universal enrichment broth was developed, enabling enrichment with multiple pathogens (69). Furthermore, other researchers have addressed this issue by developing "multiple detection capabilities," including multiplex PCR assays to detect two pathogens with one assay (34). In these protocols, identification of a specific amplicon corresponding to a specific pathogen is achieved by determining the molecular weight of the amplified product by performing agarose gel electrophoresis and visualizing the amplification product by ethidium bromide staining of agarose gels (34, 57, 156), a method not particularly suitable for large sample sizes. In order to increase the specificity of detection, a number of PCR assays utilize either scoring of the target DNA or post-PCR hybridization-capture methods (34, 38, 126, 156). These approaches have met with limited success, as these modifications make the overall pathogen detection procedure labor-intensive, time-consuming, and difficult to automate.

More recently, modifications have been introduced whereby the progress of PCR can be monitored in real time (16, 17, 142). Briefly, a reporter dye with unique excitation and emission wavelengths is introduced into the PCR mixture at concentrations which will not affect the normal functioning of the DNA polymerase. A DNA-binding dye such as SYBR green I binds double-stranded DNA and changes its emission properties when exposed to a specific wavelength of light (46, 75). The emitted fluorescence is directly proportional to the amount of DNA present in the PCR tube. The initial fluorescence due to template DNA and background emission is used to calculate the baseline data, and measurements are performed at the end of each subsequent PCR cycle. Alternatively, there are other methods to monitor the progress of PCR and thereby measure the double-stranded DNA at the end of each PCR cycle. A short oligomer fluorescent probe (30 to 50 bp in length) specific to the target gene is introduced into the PCR mixture. This probe provides additional specificity for detecting the target DNA. There are three types of probes, molecular beacon probes, TaqMan probes, and fluorescent resonance energy transfer probes, and they all possess different reporter chemistries to quantitate the target DNA (35, 44, 54, 102, 103). Collectively, real-time PCR techniques are available and each one has its unique advantages and limitations. One distinct advantage of the SYBR green I-based PCR methods is that one can modify the existing PCR protocols to suit real-time detection capability by including the dye in the PCR mix. Thus, there is minimum deviation from the original protocol and setting up validation studies or a third-party approval process is relatively easy.

Other Nucleic Acid-Based Detection Methods

Some of the alternatives to conventional PCR-based systems are RNA-based PCR using the enzyme reverse transcriptase (RT-PCR), nucleic acid sequence-based amplification, and ribotyping. One disadvantage of PCR methods is that DNA from dead pathogenic cells may be amplified during PCR and thus generate a false-positive result. Although an enrichment step prior to DNA isolation greatly reduces the chances, the possibility still remains that dead cells, if present in high numbers, may contribute to the signal intensity of PCR amplification products. However, RNA has a relatively short half-life, being rapidly degraded in dead cells. In RT-PCR, the enzyme reverse transcriptase synthesizes DNA from RNA. This DNA then serves as the template for PCR amplification. Although RT-PCR has been successfully used for detecting foodborne pathogens, no commercial kits are yet available (89, 114). Ribotyping is a robust method that can be used to generate reference data sets to identify and track specific strains associated with a production line or particular batches of raw produce. The method is based on DNA fingerprint patterns generated by specific restriction endonuclease digestion of DNA and the homology of the digested DNA fragments to the ribosomal RNA gene probes. This homology is determined by Southern blot hybridization. Individual bacterial strains carry unique ribotype patterns, and the patterns can be stored as digital images and compared to identify the source of contamination by using data from the related foodborne outbreak cases. An automated process available from Dupont can identify bacterial isolates within 16 h. Large data sets (digital images of ribotype patterns similar to fingerprints) have been generated for important foodborne pathogens such as *L. monocytogenes* and *Salmonella* spp. (62, 120).

Alternative Rapid Methods

Bioluminescence-based methods can also provide a rapid estimate of either general hygiene or residual biological materials or microbial load. The method relies on the ubiquitous presence of ATP in all living cells and the ability of ATP to emit light when in contact with the luciferase enzyme complex. Surfaces are swabbed and placed in a tube containing the reaction mixture with luciferase. ATP collected from the surface is the substrate for the luciferase. Hand-held luminometers (light-reading instruments) are available and read results as relative light units within 2 min after the swab is inserted into the luminometer chamber. The hand-held system can be used as a hygiene monitoring kit but obviously lacks the capability to detect specific pathogens.

Several chromogenic agar media have been developed for rapid detection of foodborne pathogens (18, 40). These methods may be beneficial in epidemiological investigations due to the potential to obtain results several days ahead of those obtained with the practices presently recommended by the FSIS. Accuracy and sensitivity are the two most important criteria for any new (rapid) method that is being proposed. A failure to detect the target microorganism when it is detected by the conventional method is considered a false negative, and similarly, a positive result not supported by the conventional protocol is regarded as a false positive. Methods such as real-time PCR do provide high-throughput sampling capability but require front-end capital expenditure and trained personnel. The availability of easy-to-perform, inexpensive, and rapid detection methods will directly help in monitoring of foodborne pathogens in an affordable manner.

ROLE OF REGULATORY AGENCIES IN PRODUCE SAFETY

There are several departments, agencies, and programs within the federal government that have jurisdiction over food safety issues. The Department of Health and Human Services oversees two agencies with food safety responsibility: the FDA and the CDC. The USDA houses the Agricultural Marketing Service, the Animal and Plant Health Inspection Service, the Agricultural Research Service, and the FSIS. The FSIS does not conduct any research and depends on the Agricultural Research Service and other federal and academic institutions to provide reliable scientific information. The FSIS is responsible for ensuring that meat, poultry, and egg products are safe, and the FDA is responsible for protecting consumers from adulterated and unsafe food not regulated by the FSIS. Thus, the FDA enforces its jurisdiction to monitor fresh-cut produce by requesting companies to voluntarily recall the adulterated produce or by going to court to seize the food. Both the FDA and the FSIS have embraced risk assessment and the HACCP program as foundations for regulatory approaches. As a part of the efforts to improve the safety of domestic as well as imported produce, the FDA and the USDA issued guidance to industry in October 1998. This guidance document discusses microbial food safety hazards and good agricultural practices (GAPs) and good management practices for common farm and produce operations such as growing, harvesting, washing, and packaging of minimally processed produce. The document provides a framework and guidance to both domestic and foreign fresh-fruit and -vegetable producers to help ensure the safety of produce. For a more in-depth discussion on GAPs, refer to chapter 2.

State and local government agencies are also responsible for food safety, with programs that are complementary to the federal food safety policies. States have responsibility for the consumers' health surveillance, and the local health departments are the first investigators of disease outbreaks that may be food borne. In addition, state and local governments also have authority over food service establishments that include restaurants, fast-food establishments, grocery stores, and sidewalk vendors. After the widely publicized 1993 multistate *E. coli* O157:H7 infection outbreak, the news media and professional societies have played a major role in communicating microbiological safety issues to the public and researchers. The media coverage of food recalling and disease outbreaks has increased public awareness of food safety issues. Several professional societies such as the American Society for Microbiology, the Association of Official Analytical Chemists, and the IFPA have played a valuable role in bringing scientists from different areas together to provide a multidisciplinary approach to food safety issues.

Every outbreak of foodborne illness provides an opportunity for action agencies to gain information and reassess the microbiological food safety strategies designed to ensure the safety of the food supply. Contaminated green onions have been implicated in foodborne illness outbreaks in the United States. In the fourth week of October 2003, a large outbreak of hepatitis A occurred in Pennsylvania; more than 500 people contracted the virus, and three people died (43). The estimated losses for Mexican growers totaled $10.5 million. Mexican cantaloupes were associated with a *Salmonella* infection outbreak in the United States. The FDA put all Mexican cantaloupes under import alert and certified six growers with GAPs and good management practices to export to the United States.

Two new systems were developed by the CDC to provide much-needed information about the incidence of foodborne illness: FoodNet and PulseNet. FoodNet is an active system of disease surveillance designed to provide information about the incidence of foodborne illness and to serve as the framework for case control and other epidemiological studies. FoodNet collects data on the laboratory-confirmed occurrence of bacterial pathogens such as *E. coli* O157:H7, *L. monocytogenes*, *Campylobacter*, *Salmonella*, *Shigella*, and *Yersinia enterocolitica* as well as foodborne parasites *Cyclospora* spp. and *Cryptosporidium* (3). This program covers approximately 10% of the U.S. population from nine states. The National Molecular Subtyping Network for

Foodborne Disease Surveillance, commonly referred to as PulseNet, was designed to speed the identification of outbreaks of foodborne illnesses by providing access to the molecular fingerprint data to laboratories across the United States. The FDA and FSIS laboratories can access the pulsed-field gel electrophoresis analysis data electronically to expedite identification of the pathogen in the outbreaks. Presently, data for five bacterial pathogens are included: *E. coli* O157:H7, *L. monocytogenes*, *Campylobacter*, *Salmonella*, and *Shigella*. Action agencies have used PulseNet to identify outbreaks of infection with foodborne pathogens and link them across several states, a procedure that would have taken much longer without the system. Linked to product recalls with widespread publicity, PulseNet is helping to reduce the extent of outbreaks because it initiates product recalls sooner and thus prevents more cases of illnesses.

Policymakers and researchers are concerned about why reported outbreaks associated with produce are increasing. Improved outbreak investigation tools (such as PulseNet) and better diagnostic methods have likely contributed to increased identification of cases. Research must continue on the survival and adaptation of pathogens and changes in U.S. produce consumption habits to provide a more complete view of the role of produce in human illness. For example, U.S. per capita consumption of fresh fruits and vegetables (excluding juices and other processed or canned produce) increased from 249 pounds in 1981 to 339 pounds in 2000 (104). This increase was accompanied by an increase in the variety of produce available in the market, as the number of different types of produce items in the grocery store increased from 137 in 1987 to 345 in 1998. In order to investigate food safety concerns, the USDA's Agricultural Marketing Service has implemented the microbiological data program (MDP) as part of the president's Food Safety Initiative. This nonregulatory program is expected to provide baseline information on microbial contamination of produce. The MDP determines the commodities and type of testing in consultation with the FDA, the CDC, and the National Agricultural Statistics Service. The MDP has tested cantaloupes, celery, leaf lettuce, romaine lettuce, cilantro, parsley, green onions, and tomatoes. Pathogenic *E. coli*, *Salmonella*, and *E. coli* O157:H7 are the present focus of the MDP, and methods for *Shigella* detection are expected to be released in 2006. The MDP collects samples at terminal markets and from large chain store distribution centers through eleven participating state departments of agriculture (those in California, Colorado, Florida, Maryland, Michigan, Minnesota, New York, Ohio, Texas, Washington, and Wisconsin). This sampling is projected to cover an area corresponding to approximately 51% of the U.S. population. Some concerns have been raised by the industry that since the MDP is testing at terminal markets, points of contamination at the farm or along the distribution chain cannot be determined (6). Low numbers of contaminated product at terminal markets would shift greater emphasis on control measures to the final point of service, such as food service establishments, cafeterias, and restaurants (28).

RECENT DEVELOPMENTS

Freshness Indicators

Several freshness indicators are presently being developed to indicate spoilage due to temperature abuse, package leakage, or detection of a specific compound or metabolite. If a metabolite from a microorganism (volatile amines, for example) is detected, a metabolite-specific indicator is activated and produces a visible sign to indicate that spoilage could have occurred. Some of the examples of chemical indicators are FreshTag and Toxin Guard, manufactured by Check-it ApS and Toxin Alert Inc., respectively. Similar to freshness indicators, time-temperature indicators are another type of visible indicator devices to track whether the product has suffered any time or temperature abuse. An ideal freshness or temperature indicator should be easy to include in the product package, nontoxic, and, perhaps most important, low cost. Temperature is one of the major controlling factors of food quality and food safety because of its influence on microbial growth rates. Results from several food safety studies conducted in Europe, North America, and Australia have indicated that consumers lack knowledge about adequate refrigeration practices (110, 124). Recently, the Food Hygiene Team of the National Food Research Institute, Tsukuba, Japan, developed a novel low-cost microbial sensor containing freeze-tolerant *Saccharomyces cerevisiae* for visibly detecting inappropriate temperature treatments (Fig. 8) (90). The biosensor is a simple bag containing a solution of yeast cells, yeast extract, glucose, and glycerol sealed in a multilayer transparent film with barriers against oxygen and humidity. The amount of time that food is exposed to abusive temperatures can be deduced visibly by the quantity of gas produced in the biosensor. Devices like this may offer a useful tool for securing food safety by monitoring temperature history at a very low cost during transportation, in the store, and at home.

Programs Supporting Proactive Food Safety Measures for Fresh-Cut Produce

According to the Economics Research Service of the USDA, present incentives for growers to adopt additional food safety practices are inadequate. Growers are hesitant to demand premium prices for the produce grown using stringent food safety practices (29). Interestingly, there are certain federal

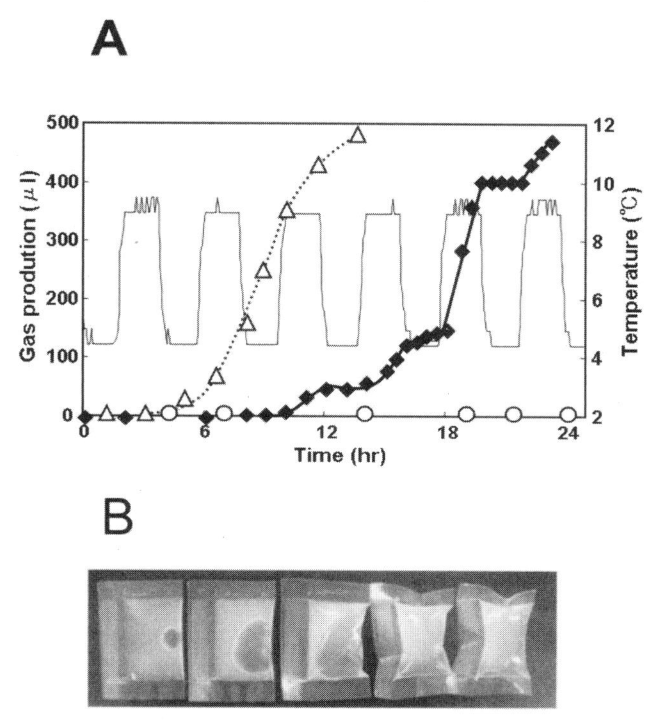

Figure 8 Gas production by biosensors. (A) Time course of gas production by a biosensor with repetition of temperature change between 5 and 10°C at 2-h intervals (◆) and with constant temperatures of 5°C (○) and 10°C (△). (B) Yeast biosensor with increasing gas production due to exposure to higher temperatures. (Reprinted from reference 90 with permission.)

programs that encourage producers to take proactive steps to implement microbial food safety objectives. For example, the USDA's "Qualified through Verification" (QTV) program is designed to verify the suitability of a firm's HACCP food safety system and to empower firms to apply science-based food safety principles to identify hazards in preparation of fresh-cut produce (148). The QTV program is a voluntary, user-fee-supported program. Under the QTV system, USDA experts work with company management to validate a facility's HACCP plan and verify the plan's effectiveness. Firms operating under the QTV program can use the USDA QTV shield on their packages, thereby displaying the added value to the product for recognition by value-oriented customers. The QTV program is presently applied only to the fresh-cut-fruit and -vegetable industry. QTV fosters a proactive approach for identifying process deficiencies during fresh-cut-produce preparations rather than after the production is completed.

Health-Promoting Compounds in Vegetables and Fruits

Many epidemiological studies have shown negative correlations between intake of fruits and vegetables and the incidence of several important diseases, including cancer and atherosclerosis (141, 149, 159). Several laboratories have taken the approach that the compounds from fruits and vegetables with established beneficial properties (for example, antioxidants) have either additional or unknown benefits and properties that have until now been overlooked. The European Union has initiated a systematic approach for identifying plant compounds that impact human health, such as falcarinol (from carrots and green tomatoes), sesquiterpene lactones (lettuce), and organosulfur compounds (onions) (21). Investigations leading to further understanding of properties and interactions of known compounds from vegetables are likely to encourage producers to identify cultivars that produce greater quantities of growth-promoting compounds.

RESOURCE INFORMATION

There is considerable information available on fresh-cut produce as well as on the microbiological safety of fresh-cut-produce preparation and storage practices. Of specific interest is the website of the IFPA (http://www.fresh-cuts.org), the University of California postharvest group website (http://postharvest.ucdavis.edu), and the USDA handbook 66 website (http://www.ba.ars.usda.gov/hb66/index.html). The National Food Research Institute of Japan has initiated an experimental website (http://seica.info) which enables consumers to trace produce purchased at retail stores to the farms of origin (138). For instance, a customer who has purchased lettuce at a supermarket which has an eight-digit SEICA catalog number would be able to trace back information such as the location of the farmland, the use of specific pesticides, and the harvest date. As of October 2004, SEICA had over 3,000 registered food products in the catalog. With increasing demand for an open information system for produce that is organically grown, Web resources such as SEICA provide an interface for producers as well as customers.

CONCLUSIONS

The phenomenal market growth in the fresh-cut-fruit and -vegetable segment is reflective of consumer trends and the need to service time-starved consumers. Preharvest contamination prevention programs and postharvest sanitation are key tools in preventing outbreaks of infection associated with fresh and fresh-cut produce. Among the greatest concerns with human pathogens on fresh-cut fruits and vegetables are enteric pathogens (e.g., *E.*

coli O157:H7 and *Salmonella*) that have the potential for growth prior to consumption, have low infective doses, or grow under refrigeration storage conditions (e.g., *L. monocytogenes*). Fresh-cut produce processors need to implement proactive programs such as the USDA QTV program and crisis management protocols to prepare for potential recalls. Although American consumers benefit from one of the world's safest food supplies, recent consumer surveys and outbreak surveillance data suggest that education efforts should continue to focus on the areas of personal and kitchen hygiene and avoidance of cross-contamination. Future changes to enhance food safety of ready-to-eat fresh-cut fruits and vegetables require a better understanding of risks associated with microbial adaptation to stress tolerance and higher economic incentives for producers to implement better sanitation strategies.

ACKNOWLEDGMENTS

I sincerely thank Robert Saftner, Shinichi Kawamoto, Shigenobu Koseki, and Jin-ichi Sugiyama for sharing their unpublished data and Manan Sharma for critically reading the manuscript.

REFERENCES

1. **Abbott, J. A., R. A. Saftner, K. C. Gross, B. T. Vinyard, and J. Janick.** 2004. Consumer evaluation and quality measurement of fresh-cut slices of "Fuji," "Golden Delicious," "Gold Rush," and "Granny Smith" apples. *Postharvest Biol. Technol.* **33**:127–140.
2. **Abe, K., N. Saito, F. Kasuga, and S. Yamamoto.** 2004. Prolonged incubation period of salmonellosis associated with low bacterial doses. *J. Food Prot.* **67**:2735–2740.
3. **Allos, B. M., M. R. Moore, P. M. Griffin, and R. V. Tauxe.** 2004. Surveillance for sporadic foodborne disease in the 21st century: the FoodNet perspective. *Clin. Infect. Dis.* **38**(Suppl. 3):S115–S120.
4. **Anonymous.** 30 January 2001, posting date. *FDA Survey of Imported Fresh Produce.* [Online.] U.S. Food and Drug Administration, Washington, D.C. http://www.cfsan.fda.gov/~dms/prodsur6.htm.
5. **Anonymous.** 2000. Fresh sliced apples: waiting to boom? *Fresh Cut* **8**:18–22.
6. **Anonymous.** 2002. USDA program to collect data on produce pathogens raises questions, concerns. *Produce News* **2002**(January 14):7.
7. **Audia, J. P., C. C. Webb, and J. W. Foster.** 2001. Breaking through the acid barrier: an orchestrated response to proton stress by enteric bacteria. *Int. J. Med. Microbiol.* **291**:97–106.
8. **Austin, J. W., K. L. Dodds, B. Blanchfield, and J. M. Farber.** 1998. Growth and toxin production by *Clostridium botulinum* on inoculated fresh-cut packaged vegetables. *J. Food Prot.* **61**:324–328.
9. **Babic, I., and A. E. Watada.** 1996. Microbial populations of fresh-cut spinach leaves affected by controlled atmospheres. *Postharvest Biol. Technol.* **9**:187–193.
10. **Bailey, J. S.** 1998. Detection of *Salmonella* cells within 24 to 26 hours in poultry samples with the polymerase chain reaction BAX system. *J. Food Prot.* **61**:792–795.

11. **Bari, M. L., M. Nakauma, S. Todoriki, V. K. Juneja, K. Isshiki, and S. Kawamoto.** 2005. Effectiveness of irradiation treatments in inactivating *Listeria monocytogenes* on fresh vegetables at refrigeration temperature. *J. Food Prot.* **68**:318–323.
12. **Basaran, N., A. Quintero-Ramos, M. M. Moake, J. J. Churey, and R. W. Worobo.** 2004. Influence of apple cultivars on inactivation of different strains of *Escherichia coli* O157:H7 in apple cider by UV irradiation. *Appl. Environ. Microbiol.* **70**:6061–6065.
13. **Berkelman, R., M. P. Doyle, D. Schaffner, and P. Hall.** 4 August 2004, posting date. *Establishing a Regulatory Limit on* Listeria monocytogenes. Docket no. 2003P-0574. [Online.] American Society for Microbiology, Washington, D.C. http://www.asm.org/Policy/index.asp?bid=29757.
14. **Bernard, R. J., and G. J. Jackson.** 1984. The transfer of human infections by foods, p. 365–378. *In* S. L. Erlandsen and E. A. Mayer (ed.), Giardia *and Giardiasis.* Plenum Press, New York, N.Y.
15. **Beuchat, L. R., and J. H. Ryu.** 1997. Produce handling and processing practices. *Emerg. Infect. Dis.* **3**:459–465.
16. **Bhagwat, A. A.** 2004. Rapid detection of *Salmonella* from vegetable rinse-water using real-time PCR. *Food Microbiol.* **21**:73–78.
17. **Bhagwat, A. A.** 2003. Simultaneous detection of *Escherichia coli* O157:H7, *Listeria monocytogenes* and *Salmonella* strains by real-time PCR. *Int. J. Food Microbiol.* **84**:217–224.
17a. **Bhagwat, A. A., L. Chan, R. Han, J. Tan, M. Kothary, J. Jean-Gilles, and B. D. Tall.** 2005. Characterization of enterohemorrhagic *Escherichia coli* strains based on acid resistance phenotypes. *Infect. Immun.* **73**:4993–5003.
18. **Bhagwat, A. A., and W. Lauer.** 2004. Food borne outbreaks in raw produce can be prevented. *Food Quality* **11**:62–63.
19. **Bhagwat, A. A., R. A. Saftner, and J. A. Abbott.** 2004. Evaluation of wash treatments for survival of foodborne pathogens and maintenance of instrumental and sensory characteristics of fresh-cut apple slices. *Food Microbiol.* **21**:319–326.
20. **Blaser, M. J., and L. S. Newman.** 1982. A review of human salmonellosis. I. Infective dose. *Rev. Infect. Dis.* **4**:1096–1106.
21. **Brandt, K., L. P. Christensen, J. Hansen-Moller, S. L. Hansen, J. Haraldsdottir, L. Jespersen, S. Purup, A. Kharazmi, V. Barkholt, H. Froker, and V. Kobak-Larsen.** 2004. Health promoting compounds in vegetables and fruits: a systematic approach for identifying plant components with impact on human health. *Trends Food Sci. Technol.* **15**:384–393.
22. **Brecht, J. K.** 1995. Physiology of lightly processed fruits and vegetables. *HortScience* **30**:18–22.
23. **Buchanan, R. L., and C. M. Deroever.** 1993. Limits in assessing microbiological food safety. *J. Food Prot.* **56**:725–729.
24. **Buck, J. W., R. R. Walcot, and L. R. Beuchat.** 21 January 2003, posting date. Recent trends in microbiological safety of fruits and vegetables. *Plant Health Prog.* [Online.] doi: 10.1094/PhP-2003-0121-01-RV.
25. **Burnett, S. L., and L. R. Beuchat.** 2001. Human pathogens associated with raw produce and unpasteurized juices, and difficulties in decontamination. *J. Ind. Microbiol. Biotechnol.* **27**:104–110.

26. **Buta, J. G., H. E. Moline, D. W. Spaulding, and C. Y. Wang.** 1999. Extending storage life of fresh-cut apples using natural products and their derivatives. *J. Agric. Food Chem.* **47**:1–6.

27. **Buzby, J. C., and J. F. Guthrie.** 2004. *The USDA Fruit and Vegetable Pilot Program Evaluation.* Agriculture information bulletin 792-6, October 2004. Economic Research Service, U.S. Department of Agriculture, Washington, D.C.

28. **Calvin, L.** 2003. *Response to U.S. Foodborne Illness Outbreaks Associated with Imported Produce.* Report 828. Economic Research Service, U.S. Department of Agriculture, Washington, D.C.

29. **Calvin, L., B. Avendano, and R. Schwentesius.** 2004. *The Economics of Food Safety: the Case of Green Onion and Hepatitis A Outbreaks.* Agricultural economic report number 305. Economic Research Service, U.S. Department of Agriculture, Washington, D.C.

30. **Campbell, J. V., J. Mohle-Boetani, R. Reporter, S. Abbott, J. Farrar, M. Brandl, R. Mandrell, and S. B. Werner.** 2001. An outbreak of *Salmonella* serotype *Thompson* associated with fresh cilantro. *J. Infect. Dis.* **183**:984–987.

31. **Castanie-Cornet, M. P., T. A. Penfound, D. Smith, J. F. Elliott, and J. W. Foster.** 1999. Control of acid resistance in *Escherichia coli. J. Bacteriol.* **181**:3525–3535.

32. **Centers for Disease Control and Prevention.** 1991. Multistate outbreak of *Salmonella poona* infections—United States and Canada. *Morb. Mortal. Wkly. Rep.* **40**:549–552.

33. **Centers for Disease Control and Prevention.** 2004. Preliminary FoodNet data on the incidence of infection with pathogens transmitted commonly through food—selected sites, United States, 2003. *Morb. Mortal. Wkly. Rep.* **53**:338–343.

34. **Chen, J., and M. W. Griffiths.** 2001. Detection of *Salmonella* and simultaneous detection of *Salmonella* and Shiga-like toxin-producing *Escherichia coli* using the magnetic capture hybridization polymerase chain reaction. *Lett. Appl. Microbiol.* **32**:7–11.

35. **Chen, W., G. Martinez, and A. Mulchandani.** 2000. Molecular beacons: a real-time polymerase chain reaction assay for detecting *Salmonella. Anal. Biochem.* **280**:166–172.

36. **Chen, Y., W. H. Ross, V. N. Scott, and D. E. Gombas.** 2003. *Listeria monocytogenes*: low levels equals low risk. *J. Food Prot.* **66**:570–577.

37. **Cheville, A. M., K. W. Arnold, C. Buchrieser, C. M. Cheng, and C. W. Kaspar.** 1996. *rpoS* regulation of acid, heat, and salt tolerance in *Escherichia coli* O157:H7. *Appl. Environ. Microbiol.* **62**:1822–1824.

38. **Cocolin, L., M. Manzano, C. Cantoni, and G. Comi.** 1998. Use of polymerase chain reaction and restriction enzyme analysis to directly detect and identify *Salmonella typhimurium* in food. *J. Appl. Microbiol.* **85**:673–677.

39. **Code of Federal Regulations.** 1 April 2000, posting date. *Title 21: Food and Drugs,* section 101. 95, p. 145–146. [Online.] U.S. Government Printing Office, Washington, D.C. http://frwebgate.access.gpo.gov/cgi-bin/get-cfr.cgi?TITLE=21&PART=101&SECTION=95&TYPE=TEXT&YEAR=2000.

40. **Cooke, V. M., R. J. Miles, R. G. Price, and A. C. Richardson.** 1999. A novel chromogenic ester agar medium for detection of salmonellae. *Appl. Environ. Microbiol.* **65**:807–812.

41. **D'Aoust, J. Y.** 1985. Infective dose of *Salmonella typhimurium* in cheddar cheese. *Am. J. Epidemiol.* **122**:717–719.

42. **D'Aoust, J. Y., B. J. Aris, P. Thisdele, A. Durante, N. Brisson, D. Dragon, G. Leachapelle, M. Johnston, and R. Laidley.** 1975. *Salmonella eastbourne* outbreak associated with chocolate. *Can. Inst. Food. Sci. Technol. J.* **8:**181–184.
43. **Dato, V., A. Weltman, K. Waller, M. Ruta, A. Highbaugh-Battle, C. Hembree, S. Evenson, C. Wheeler, and T. Vogt.** 2003. Hepatitis A outbreak associated with green onion at a restaurant—Monaca, Pennsylvania, 2003. *Morb. Mortal. Wkly. Rep.* **52:**1155–1157.
44. **Daum, L. T., W. J. Barnes, J. C. McAvin, M. S. Neidert, L. A. Cooper, W. B. Huff, L. Gaul, W. S. Riggins, S. Morris, A. Salem, and K. L. Lohman.** 2002. Real-time detection of *Salmonella* in suspect foods from a gastroenteritis outbreak in Kerr County, Texas. *J. Clin. Microbiol.* **40:**3050–3052.
45. **Davidson, P. M., and V. K. Juneja.** 1990. Antimicrobial agents, p. 83–137. *In* A. L. Branen, P. M. Davidson, and S. Salminen (ed.), *Food Additives*. Marcel Dekker, New York, N.Y.
46. **De Medici, D., L. Croci, E. Delibato, S. Di Pasquale, E. Filetici, and L. Toti.** 2003. Evaluation of DNA extraction methods for use in combination with SYBR Green I real-time PCR to detect *Salmonella enterica* serotype Enteritidis in poultry. *Appl. Environ. Microbiol.* **69:**3456–3461.
47. **Donnelly, C. W.** 2001. *Listeria monocytogenes,* p. 99–132. *In* R. Labbe and S. Garcia (ed.), *Guide to Foodborne Pathogens*. John Wiley & Sons, Inc., New York, N.Y.
48. **Duffy, E. A., L. M. Lucia, J. M. Kells, A. Castillo, S. D. Pillai, and G. R. Acuff.** 2005. Concentration of *Escherichia coli* and genetic diversity and antibiotic resistance profiling of *Salmonella* isolated from irrigation water, packing shed equipment, and fresh produce in Texas. *J. Food Prot.* **68:**70–79.
49. **Fan, X., P. M. Toivonen, K. T. Rajkowski, and K. J. Sokorai.** 2003. Warm water treatment in combination with modified atmosphere packaging reduces undesirable effects of irradiation on the quality of fresh-cut iceberg lettuce. *J. Agric. Food Chem.* **51:**1231–1236.
50. **Farber, J. M., and P. I. Peterkin.** 1991. *Listeria monocytogenes*, a food-borne pathogen. *Microbiol. Rev.* **55:**476–511.
51. **Feng, P.** 1997. Impact of molecular biology on the detection of foodborne pathogens. *Mol. Biotechnol.* **7:**267–278.
52. **Ferenci, T.** 2003. What is driving the acquisition of *mutS* and *rpoS* polymorphisms in *Escherichia coli*? *Trends Microbiol.* **11:**457–461.
53. **Foegeding, P. M., V. Hemstapat, and F. G. Giesbrecht.** 1986. Chlorine dioxide inactivation of *Bacillus* and *Clostridium* spores. *J. Food Sci.* **51:**197–201.
54. **Fortin, N. Y., A. Mulchandani, and W. Chen.** 2001. Use of real-time polymerase chain reaction and molecular beacon for the detection of *Escherichia coli* O157:H7. *Anal. Biochem.* **289:**281–288.
55. **Foschino, R., I. Nervegena, A. Motta, and A. Galli.** 1998. Bactericidal activity of chlorine dioxide against *Escherichia coli* in water and on hard surfaces. *J. Food Prot.* **61:**668–672.
56. **Francis, G. A., C. Thomas, and D. O'Beirne.** 1999. The microbiological safety of minimally processed vegetables. *Int. J. Food Sci. Technol.* **34:**1–22.
57. **Fratamico, P., and T. P. Strobaugh.** 1998. Simultaneous detection of *Salmonella* spp. and *Escherichia coli* O157:H7 by multiplex PCR. *J. Ind. Microbiol. Biotechnol.* **21:**92–98.
58. **Frenzen, P. D.** 2004. Deaths due to unknown foodborne agents. *Emerg. Infect. Dis.* **10:**1536–1543.

59. **Frenzen, P. D., A. Majchrowicz, J. C. Buzby, and B. Imhoff.** 2000. Consumer acceptance of irradiated meat and poultry products. *Agric. Information Bull.* **757**:1–8.
60. **Garcia, E., and D. M. Barrett.** 2002. Preservative treatment for fresh-cut fruits and vegetables, p. 267–303. *In* O. Lamikarira (ed.), *Fresh-Cut Fruits and Vegetables: Science, Technology and Market.* CRC Press, Boca Raton, Fla.
61. **Gawande, P. V., and A. A. Bhagwat.** 2002. Inoculation onto solid surfaces protects *Salmonella* spp. during acid challenge: a model study using polyethersulfone membranes. *Appl. Environ. Microbiol.* **68**:86–92.
62. **Gendel, S. M., and J. Ulaszek.** 2000. Ribotype analysis of strain distribution in *Listeria monocytogenes. J. Food Prot.* **63**:179–185.
63. **Gombas, D. E., Y. Chen, R. S. Clavero, and V. N. Scott.** 2003. Survey of *Listeria monocytogenes* in ready-to-eat foods. *J. Food Prot.* **66**:559–569.
64. **Gorny, J. R. (ed.).** 2001. *Food Safety Guidelines for the Fresh-Cut Produce Industry,* 4th ed. International Fresh-Cut Produce Association, Alexandria, Va.
65. **Gorny, J. R.** 2003. New opportunities for fresh-cut apples. *Fresh Cut* **11**:14–15.
66. **Gorny, J. R.** 1997. A summary of CA and MA requirements and recommendations for fresh-cut (minimally processed) fruits and vegetables, p. 30–66. *In* J. R. Gorny (ed.), *Proceedings of the 7th International Controlled Atmosphere Research Conference,* vol. 5. University of California, Davis, Davis, Calif.
67. **Goulet, V., H. de Valk, O. Pierre, F. Stainer, J. Rocourt, V. Vaillant, C. Jacquet, and J. Desenclos.** 2001. Effect of prevention measures on incidence of human listeriosis, France, 1987–1997. *Emerg. Infect. Dis.* **7**:983–989.
68. **Gunes, G., and C. Y. Lee.** 1997. Color of minimally processed potatoes as affected by modified atmosphere packaging and anti-browning agents. *J. Food Sci.* **62**:572–575.
69. **Hammack, T. S., R. M. Amaguana, and W. H. Andrews.** 2001. An improved method for the recovery of *Salmonella* serovars from orange juice using universal preenrichment broth. *J. Food Prot.* **64**:659–663.
70. **Han, Y., A. M. Guentert, R. S. Smith, R. H. Linton, and R. E. Nelson.** 1999. Efficacy of chlorine dioxide gas as a sanitizer for tanks used for aseptic juice storage. *Food Microbiol.* **16**:53–61.
71. **Han, Y., R. H. Linton, S. S. Nielsen, and P. E. Nelson.** 2000. Inactivation of *Escherichia coli* O157:H7 on surface-uninjured and -injured green pepper (*Capsicum annuum* L.) by chlorine dioxide gas as demonstrated by confocal laser scanning microscopy. *Food Microbiol.* **17**:643–655.
72. **Harp, E., and S. E. Gilliland.** 2003. Evaluation of a select strain of *Lactobacillus delbrueckii* subsp. *lactis* as a biological control agent for pathogens on fresh-cut vegetables stored at 7°C. *J. Food Prot.* **66**:1013–1018.
73. **Harris, L. J., J. N. Farber, L. R. Beuchat, M. E. Parish, T. V. Suslow, E. H. Garret, and F. F. Busta.** 2003. *Comprehensive Reviews in Food Science and Food Safety,* vol. 2, p. 78–89. Institute of Food Technology, Chicago, Ill.
74. **Hedberg, C. W., J. A. Korlath, J. Y. D'Aoust, K. E. White, W. L. Schell, M. R. Miller, D. N. Cameron, K. L. MacDonald, and M. T. Osterholm.** 1992. A multistate outbreak of *Salmonella javiana* and *Salmonella oranienburg* infections due to consumption of contaminated cheese. *JAMA* **268**:3203–3240.
75. **Higuchi, R., C. Fockler, G. Dollinger, and R. Watson.** 1993. Kinetic PCR analysis: real-time monitoring of DNA amplification reactions. *Bio/Technology* **11**:1026–1030.

76. **Hilborn, E. D., J. H. Mermin, P. A. Mshar, J. L. Hadler, A. Voetsch, C. Wojtkunski, M. Swartz, R. Mshar, M. A. Lambert-Fair, J. A. Farrar, M. K. Glynn, and L. Slutsker.** 1999. A multistate outbreak of *Escherichia coli* O157:H7 infections associated with consumption of mesclun lettuce. *Arch. Intern. Med.* **159**:1758–1764.

77. **Hilgren, J. D., and J. A. Salverda.** 2000. Antimicrobial efficacy of a peroxyacetic/octanoic acid mixture in fresh-cut-vegetable process waters. *J. Food Sci.* **65**:1376–1379.

78. **Hill, W. E.** 1996. The polymerase chain reaction: applications for the detection of foodborne pathogens. *Crit. Rev. Food Sci. Nutr.* **36**:123–173.

79. **Hines, E.** 2000. PCR-based testing: unraveling the mystery. *Food Quality* **7**:22–28.

80. **Hong, J. H., and K. C. Gross.** 2001. Maintaining quality of fresh-cut tomato slices through modified atmosphere packaging and low temperature storage. *J. Food Sci.* **66**:960–965.

81. **Hudson, J. A., C. Billington, G. Carey-Smith, and G. Greening.** 2005. Bacteriophages as biocontrol agents in food. *J. Food Prot.* **68**:426–437.

82. **Ibanez-Ruiz, M., V. Robbe-Saule, D. Hermant, S. Labrude, and F. Norel.** 2000. Identification of RpoS (σ^s) -regulated genes in *Salmonella enterica* serovar Typhimurium. *J. Bacteriol.* **182**:5749–5756.

83. **Jones, T., B. Imhoff, M. Samuel, and P. Mshar.** 2004. Limitations to successful investigation and reporting of foodborne outbreaks: an analysis of foodborne disease outbreaks in FoodNet catchment areas, 1998–1999. *Clin. Infect. Dis.* **38**(Suppl. 3):S297–S302.

84. **Joseph, B., S. K. Otta, I. Karunasagar, and I. Karunasagar.** 2001. Biofilm formation by *Salmonella* spp. on food contact surfaces and their sensitivity to sanitizers. *Int. J. Food Microbiol.* **64**:367–372.

85. **Kader, A. A., D. Zagory, and E. L. Kerbel.** 1989. Modified atmosphere packaging of fruits and vegetables. *Crit. Rev. Food Sci. Nutr.* **28**:1–30.

86. **Karaibrahimoglu, Y., X. Fan, G. M. Sapers, and K. Sokorai.** 2004. Effect of pH on the survival of *Listeria innocua* in calcium ascorbate solutions and on quality of fresh-cut apples. *J. Food Prot.* **67**:751–757.

87. **Kim, J. G., Y. Luo, R. A. Saftner, and K. C. Gross.** 2005. Delayed modified atmosphere packaging of fresh-cut romaine lettuce: effects on quality maintenance and shelf-life. *J. Am. Soc. Hortic. Sci.* **130**:116–123.

88. **King, T., A. Ishihama, A. Kori, and T. Ferenci.** 2004. A regulatory trade-off as a source of strain variation in the species *Escherichia coli. J. Bacteriol.* **186**:5614–5620.

89. **Klein, P. G., and V. K. Juneja.** 1997. Sensitive detection of viable *Listeria monocytogenes* by reverse transcriptase-PCR. *Appl. Environ. Microbiol.* **63**:4441–4448.

90. **Kogure, H., S. Kawasaki, K. Nakajima, N. Sakai, K. Futase, Y. Inatsu, M. L. Bari, K. Isshiki, and S. Kawamoto.** 2005. Development of a novel microbial sensor with Baker's yeast cells for monitoring temperature control during cold food chain. *J. Food Prot.* **68**: 182–186.

91. **Koseki, S., and K. Itoh.** 2002. Effect of nitrogen gas packaging on the quality and microbial growth of fresh-cut vegetables under low temperatures. *J. Food Prot.* **65**:326–332.

92. **Kumar, V., D. J. Mills, J. D. Anderson, and A. K. Mattoo.** 2004. An alternative agriculture system is defined by a distinct expression profile of select gene transcripts and proteins. *Proc. Natl. Acad. Sci. USA* **101**:10535–10540.

93. **Kusumaningrum, H. D., E. D. van Asselt, R. R. Beumer, and M. H. Zwietering.** 2004. A quantitative analysis of cross-contamination of *Salmonella* and *Campylobacter* spp. via domestic kitchen surfaces. *J. Food Prot.* **67**:1892–1903.

94. **LeBlanc, D. I., R. Stark, B. MacNeil, B. Goguen, and C. Beaulieu.** 1996. *New Developments in Refrigeration and Food Safety and Quality,* p. 42–51. International Institute of Refrigeration, Paris, France.

95. **Lee, S.-Y., M. Costello, and D.-H. Kang.** 2004. Efficacy of chlorine dioxide gas as a sanitizer of lettuce leaves. *J. Food Prot.* **67**:1371–1376.

96. **Leistner, L., and G. Gorris.** 1995. Food preservation by hurdle technology. *Trends Food Sci. Technol.* **6**:41–46.

97. **Leverentz, B., W. S. Conway, M. J. Camp, W. J. Janisiewicz, T. Abuladze, M. Yang, R. A. Saftner, and A. Sulakvelidze.** 2003. Biocontrol of *Listeria monocytogenes* on fresh-cut produce by treatment with lytic bacteriophages and a bacteriocin. *Appl. Environ. Microbiol.* **69**:4519–4526.

98. **Leverentz, B., W. S. Conway, W. J. Janisiewicz, and M. J. Camp.** 2004. Optimizing concentration and timing of a phage spray application to reduce *Listeria monocytogenes* on honeydew melon tissue. *J. Food Prot.* **67**:1682–1686.

99. **Li, C.-M., S. S. Moon, M. P. Doyle, and K. H. McWatters.** 2002. Inactivation of *Escherichia coli* O157:H7, *Salmonella enterica* serotype Enteritidis, and *Listeria monocytogenes* on lettuce by hydrogen peroxide and lactic acid and by hydrogen peroxide with mild heat. *J. Food Prot.* **65**:1215–1220.

100. **Liao, C.-H., and P. H. Cooke.** 2001. Response to trisodium phosphate treatment of *Salmonella* Chester attached to fresh-cut green pepper slices. *Can. J. Microbiol.* **47**:25–32.

101. **Liao, C. H., and G. M. Sapers.** 2000. Attachment and growth of *Salmonella chester* on apple fruits and in vivo response of attached bacteria to sanitizer treatments. *J. Food Prot.* **63**:876–883.

102. **Liming, S. H., and A. A. Bhagwat.** 2004. Application of a molecular beacon–real-time PCR technology to detect *Salmonella* species contaminating fruits and vegetables. *Int. J. Food Microbiol.* **95:177–187.**

103. **Liming, S. H., Y. Zhang, J. Meng, and A. A. Bhagwat.** 2004. Detection of *Listeria monocytogenes* in fresh produce using molecular beacon–real-time PCR technology. *J. Food Sci.* **69**:240–245.

104. **Lin, B. H.** 2004. *Fruits and Vegetables Consumption: Looking Ahead to 2010.* Report 792-7. Economic Research Service, U.S. Department of Agriculture, Washington, D.C.

105. **Lin, B. H., J. Reed, and G. Lucier.** 2004. *U.S. Fruit and Vegetable Consumption: Who, What, Where, and How Much?* Report 792-2. Economic Research Service, U.S. Department of Agriculture, Washington, D.C.

106. **Lin, J., M. P. Smith, K. C. Chapin, H. S. Baik, G. N. Bennett, and J. W. Foster.** 1996. Mechanisms of acid resistance in enterohemorrhagic *Escherichia coli. Appl. Environ. Microbiol.* **62**:3094–3100.

107. **Loewe, L., V. Textor, and S. Scherer.** 2003. High deleterious genomic mutation rate in stationary phase of *Escherichia coli. Science* **302**:1558–1560.

108. **Lombardo, M., I. Aponyi, and S. M. Rosenberg.** 2004. General stress response regulator RpoS in adaptive mutation and amplification in *Escherichia coli. Genetics* **166**:669–680.

109. **Luo, Y., J. L. McEvoy, M. R. Wachtel, J. G. Kim, and Y. Huang.** 2004. Package atmosphere affects postharvest biology and quality of fresh-cut cilantro leaves. *HortScience* **39:**567–570.

110. **Marklinder, I. M., M. Lindbald, L. M. Eriksson, A. M. Finnson, and R. Lindqvist.** 2004. Home storage temperatures and consumer handling of refrigerated foods in Sweden. *J. Food Prot.* **67:**2570–2577.

111. **McCullough, N. B., and C. W. Eisele.** 1951. Experimental human salmonellosis. I. Pathogenicity of strains of *Salmonella meleagridis* and *Salmonella anatum* obtained from spray-dried whole egg. *J. Infect. Dis.* **88:**278–289.

112. **McWatters, L. H., M. S. Chinnan, S. L. Walker, M. P. Doyle, and C. M. Lin.** 2002. Consumer acceptance of fresh-cut iceberg lettuce treated with 2% hydrogen peroxide and mild heat. *J. Food Prot.* **65:**1221–1226.

113. **Mohle-Boetani, J. C., R. Reporter, S. B. Werner, S. Abbott, J. Farrar, S. H. Waterman, and D. J. Vugia.** 1999. An outbreak of *Salmonella* serogroup saphra due to cantaloupes from Mexico. *J. Infect. Dis.* **180:**1361–1364.

114. **Morin, N. J., Z. Gong, and X. F. Li.** 2004. Reverse transcription-multiplex PCR assay for simultaneous detection of *Escherichia coli* O157:H7, *Vibrio cholerae* O1, and *Salmonella Typhi. Clin. Chem.* **50:**2037–2044.

115. **National Advisory Committee on Microbiological Criteria for Foods.** 1999. Microbiological safety evaluations and recommendations on fresh produce. *Food Control* **10:** 117–143.

116. **Nguyen-the, C., and F. Carlin.** 1994. The microbiology of minimally processed fresh fruit and vegetables. *Crit. Rev. Food Sci. Nutr.* **34:**371–401.

117. **Notley-McRobb, L., T. King, and T. Ferenci.** 2002. *rpoS* mutations and loss of general stress resistance in *Escherichia coli* populations as a consequence of conflict between competing stress responses. *J. Bacteriol.* **184:**806–811.

118. **Nwachuka, N., and C. P. Gerba.** 2004. Microbial risk assessment: don't forget the children. *Curr. Opin. Microbiol.* **7:**206–209.

119. **Nystrom, T.** 2004. Stationary-phase physiology. *Annu. Rev. Microbiol.* **58:**161–181.

120. **Oscar, T. P.** 1998. Identification and characterization of *Salmonella* isolates by automated ribotyping. *J. Food Prot.* **61:**519–524.

121. **Parnell, T. L., L. J. Harris, and T. V. Suslow.** Reducing *Salmonella* on cantaloupes and honeydew melons using wash practices applicable to postharvest handling, foodservice, and consumer preparation. *Int. J. Food Microbiol.* **99:**59–70.

122. **Public Health Laboratory Systems.** 2001. *Salmonella* Newport infection in England associated with the consumption of ready to eat salad. *Eurosurveillance Wkly.* **2001** (28 June):010628.

123. **Rayner, J., R. Veeh, and J. Flood.** 2004. Prevalence of microbial biofilms on selected fresh produce and household surfaces. *Int. J. Food Microbiol.* **95:**29–39.

124. **Redmond, E. C., and C. J. Griffith.** 2003. Consumer food handling in the home: a review of food safety studies. *J. Food Prot.* **66:**130–161.

125. **Richards, G. M., J. W. Buck, and L. R. Beuchat.** 2004. Survey of yeasts for antagonistic activity against *Salmonella* Poona in cantaloupe juice and wounds in rinds co-infected with phytopathogenic molds. *J. Food Prot.* **67:**2132–2142.

126. **Rijpens, N., L. Herman, F. Vereecken, G. Jannes, J. D. Smedt, and L. D. Zutter.** 1999. Rapid detection of stressed *Salmonella* spp. in dairy and egg products using immunomagnetic separation and PCR. *Int. J. Food Microbiol.* **46:**37–44.
127. **Saltveit, M. E.** 2003. Effect of 1–methylcyclopropane on phenylpropanoid metabolism, the accumulation of phenolic compounds, and browning of whole and fresh-cut "iceberg" lettuce. *Postharvest Biol. Technol.* **34:**75–80.
128. **Sapers, G. M.** 2003. Washing and sanitizing raw materials for minimally processed fruits and vegetable products, p. 221–253. *In* J. S. Novak, G. M. Sapers, and V. K. Juneja (ed.), *Microbial Safety of Minimally Processed Foods.* CRC Press, Washington, D.C.
129. **Sapers, G. M., L. Garzarella, and V. Pilizota.** 1990. Application of browning inhibitors to cut apple and potato by vacuum and pressure infiltration. *J. Food Sci.* **55:**1049–1053.
130. **Sapers, G. M., R. L. Miller, V. Pilizota, and A. M. Mattrazzo.** 2001. Antimicrobial treatments for minimally processed cantaloupe melon. *J. Food Sci.* **66:**345–349.
131. **Sewell, A. M., and J. M. Farber.** 2001. Foodborne outbreaks in Canada linked to produce. *J. Food Prot.* **64:**1863–1877.
132. **Sharma, M., and L. R. Beuchat.** 2004. Sensitivity of *Escherichia coli* O157:H7 to commercially available alkaline cleaners and subsequent resistance to heat and sanitizers. *Appl. Environ. Microbiol.* **70:**1795–1803.
133. **Silvestro, L., M. Caputo, S. Blancato, L. Decastelli, A. Fioravanti, R. Tozzoli, S. Morabito, and A. Caprioli.** 2004. Asymptomatic carriage of verocytotoxin-producing *Escherichia coli* O157:H7 in farm workers in Northern Italy. *Epidemiol. Infect.* **132:**915–919.
134. **Smith, A. B., J. Song, and A. C. Cameron.** 1998. Modified atmosphere packaged cut iceberg: effect of temperature and O_2 partial pressure on respiration and quality. *J. Agric. Food Chem.* **46:**4556–4562.
135. **Soliva-Fortuny, R. C., and O. Martin-Belloso.** 2003. New advances in extending the shelf-life of fresh-cut fruits: a review. *Trends Food Sci. Technol.* **14:**341–353.
136. **Solomon, H. M., D. A. Kautter, T. Lilly, and E. J. Rhodehame.** 1990. Outgrowth of *Clostridium botulinum* in shredded cabbage at room temperature under modified atmosphere. *J. Food Prot.* **53:**831–833.
137. **Steele, M., and J. Odumeru.** 2004. Irrigation water as source of foodborne pathogens on fruits and vegetables. *J. Food Prot.* **67:**2839–2849.
138. **Sugiyama, J.** 2004. "SEICA" accountability system for fruits and vegetables, p. 101–103. *In* J. P. Cherry and A. E. Pavlath (ed.), *US-Japan Cooperative Program in Natural Resources.* USDA Agricultural Research Service, Honolulu, Hawaii.
139. **Suslow, T. V., M. P. Oria, L. R. Beuchat, E. H. Garret, M. E. Parish, L. J. Harris, J. N. Farber, and F. F. Busta.** 2003. Production practices as risk factors in microbial food safety of fresh and fresh-cut produce. *Comp. Rev. Food Sci. Food Safety* **2**(Suppl.):38–72.
140. **Thomas, L. V., M. R. Clarkson, and J. Delves-Broughton.** 2000. Nisin, p. 463–524. *In* A. S. Naidu (ed.), *Natural Food Antimicrobial Systems.* CRC Press, Boca Raton, Fla.
141. **Trichopoulou, A., A. Naska, A. Antoniou, S. Friel, K. Trygg, and A. Turrini.** 2003. Vegetables and fruits: the evidence in their favor and the public health perspective. *Int. J. Vitamin Nutr. Res.* **73:**63–69.

142. **Tyagi, S., and F. R. Kramer.** 1996. Molecular beacon: probes that fluoresce upon hybridization. *Nat. Biotechnol.* **14**:303–308.

143. **Ukuku, D. O.** 2004. Effect of hydrogen peroxide treatment on microbial quality and appearance of whole and fresh-cut melons contaminated with *Salmonella* spp. *Int. J. Food Microbiol.* **95**:137–146.

144. **Ukuku, D. O., and W. F. Fett.** 2004. Effect of nisin in combination with EDTA, sodium lactate, and potassium sorbate for reducing *Salmonella* on whole and fresh-cut cantaloupe. *J. Food Prot.* **67**:2143–2150.

145. **Ukuku, D. O., V. Pilizota, and G. M. Sapers.** 2004. Effect of hot water and hydrogen peroxide treatments on survival of *Salmonella* and microbial quality of whole and fresh-cut cantaloupe. *J. Food Prot.* **67**:432–437.

146. **USDA Agricultural Marketing Service.** 1998. *Shipping Point and Market Inspections Instructions for Fresh-Cut Produce.* Fruits and Vegetable Division, Fresh Products Branch, U.S. Department of Agriculture, North Highlands, Calif.

147. **USDA Agricultureal Research Service.** 30 August 2005, posting date. *National Nutrient Database for Standard Reference.* Release 17. [Online.] USDA Agricultural Research Service, Beltsville, Md. http://www.nal.usda.gov/fnic/foodcomp.

148. **U.S. Department of Agriculture.** 2002. *"Qualified Through Verification" (QTV) Program for the Fresh-Cut Produce Industry.* Agricultural Marketing Service file code 151-B-4. U.S. Department of Agriculture, Washington, D.C.

149. **van'tVeer, P., M. C. J. F. Jansen, M. Klerk, and F. J. Kok.** 2000. Fruits and vegetables in prevention of cancer and cardiovascular disease. *Public Health Nutr.* **3**:103–107.

150. **Venkitanarayanan, K. S., C. M. Lin, H. Bailey, and M. P. Doyle.** 2002. Inactivation of *Escherichia coli* O157:H7, *Salmonella* Enteritidis, and *Listeria monocytogenes* on apples, oranges, and tomatoes by lactic acid with hydrogen peroxide. *J. Food Prot.* **65**:100–105.

151. **Wachtel, M. R., and A. O. Charkowski.** 2002. Cross-contamination of lettuce with *Escherichia coli* O157:H7. *J. Food Prot.* **65**:465–470.

152. **Wachtel, M. R., L. C. Whitehead, and R. E. Mandrell.** 2002. Association of *Escherichia coli* O157:H7 with preharvest leaf lettuce upon exposure to contaminated irrigation water. *J. Food Prot.* **65**:18–25.

153. **Wang, C. Y.** 1998. Methyl jasmonate inhibits postharvest sprouting and improves storage quality of radishes. *Postharvest Biol. Technol.* **14**:179–183.

154. **Waterman, S. R., and P. L. C. Small.** 1998. Acid-sensitive enteric pathogens are protected from killing under extremely acidic conditions of pH 2.5 when they are inoculated onto certain solid food sources. *Appl. Environ. Microbiol.* **64**:3882–3886.

155. **Waterman, S. R., and P. L. C. Small.** 1996. Characterization of the acid resistance phenotype and *rpoS* alleles of Shiga-like toxin-producing *Escherichia coli*. *Infect. Immun.* **64**:2808–2811.

156. **Weagant, S. D., J. A. Jagow, K. C. Jinneman, C. J. Omiecinski, C. A. Kaysner, and W. E. Hill.** 1999. Development of digoxigenin-labeled PCR amplicon probes for use in detection and identification of enteropathogenic *Yersinia* and Shiga toxin-producing *Escherichia coli* from foods. *J. Food Prot.* **62**:438–443.

157. **Wilcock, A., M. Pun, J. Khanona, and M. Aung.** 2004. Consumer attitudes, knowledge and behaviour: a review of food safety issues. *Trends Food Sci. Technol.* **15**:56–66.

158. **Woteki, C. E., and B. D. Kineman.** 2003. Challenges and approaches to reducing foodborne illness. *Annu. Rev. Nutr.* **23**:315–344.

159. **Wright, M. E., S. T. Mayne, C. A. Swanson, R. Sinha, and M. C. R. Alavanja.** 2003. Dietary carotenoids, vegetables, and lung cancer risk in women: the Missouri women's health study (United States). *Cancer Causes Control* **14**:85–96.

160. **Yucan, J. T. C.** 2003. Modified atmosphere packaging for shelf-life extension, p. 205-219. *In* J. S. Novak, G. M. Sapers, and V. K. Juneja (ed.), *Microbial Safety of Minimally Processed Foods.* CRC Press, Boca Raton, Fla.

161. **Zhao, C., B. Ge, J. D. Villena, R. Sudler, E. Yeh, S. Zhao, D. G. White, D. Wagner, and J. Meng.** 2001. Prevalence of *Campylobacter* spp., *Escherichia coli*, and *Salmonella* serovars in retail chicken, turkey, pork, and beef from the greater Washington, D.C., area. *Appl. Environ. Microbiol.* **67**:5431–5436.

162. **Zhuang, H., M. M. Barth, and T. R. Hankinson.** 2003. Microbial safety, quality, and sensory aspects of fresh-cut fruits and vegetables, p. 255–278. *In* J. S. Novak, G. M. Sapers, and V. K. Juneja (ed.), *Microbial Safety of Minimally Processed Foods.* CRC Press, Boca Raton, Fla.

Microbiology of Fresh Produce
Edited by Karl R. Matthews

Seed Sprouts: the State of Microbiological Safety

6

William F. Fett, Tong-Jen Fu, and Mary Lou Tortorello

Mung bean sprouts have been propagated and used as food by the Chinese for almost 5,000 years (132). Presently, a wide variety of seed sprouts are grown at commercial establishments or in the home for consumption either raw or lightly cooked. In the United States, mung bean, alfalfa, clover, radish, and broccoli are among the most popular sprouts. Sprouts are propagated primarily by placing the seeds in trays, rotary drums, or bins and watering frequently for 4 to 7 days. Soil and soil-less planting mixes are used for only a few types of sprouts, including sunflower and wheatgrass, which are sometimes grown in greenhouses. Due to limited shelf life after harvest, packaged sprouts are distributed either locally or regionally.

Sprouts are believed by many consumers to be a healthy natural food. The health benefits of sprout consumption, in addition to the nutritional value of sprouts, are being confirmed by recent research efforts in several laboratories around the world. For example, the consumption of broccoli sprouts may provide chemoprotection against certain carcinogens as well as reduce the risk of developing hypertension and atherosclerosis (35, 190). Among the numerous tested vegetables, alfalfa sprouts ranked near the top in antioxidant activity (21).

Unfortunately, in the past decade, consumption of raw or lightly cooked seed sprouts contaminated with *Salmonella enterica* or *Escherichia coli* O157:H7 or O157:NM has been responsible for at least 27 outbreaks of infection in the United States, resulting in over 1,600 reported cases of

WILLIAM F. FETT, Food Safety Intervention Technologies Research Unit, Eastern Regional Research Center, Agricultural Research Service, U.S. Department of Agriculture, Wyndmoor, PA 19038. TONG-JEN FU AND MARY LOU TORTORELLO, National Center for Food Safety and Technology, U.S. Food and Drug Administration, Summit-Argo, IL 60501.

foodborne illness (54). During the period from 1973 through 1997, contaminated seed sprouts were responsible for 7 of 30 recorded U.S. outbreaks of foodborne salmonellosis linked to contaminated produce. Furthermore, contaminated sprouts were responsible for more multistate foodborne outbreaks than any other single produce item (154). Due to the international nature of seed distribution systems, some of these outbreaks have been international in scope (135). Sprouts were designated as a special food safety problem by the National Advisory Committee on Microbiological Criteria for Food in 1999 (117) for two reasons: (i) low levels of human pathogens on sprout seeds multiply into high populations during the sprouting process due to favorable conditions of temperature, moisture, and nutrient availability, and (ii) seed sprouts are often consumed raw, with no killing step. Sprouts are referred to as a "potentially hazardous food" in the 2001 U.S. Food and Drug Administration (FDA) Food Code (51). The FDA has issued a number of consumer advisories informing the public about the risks associated with eating raw sprouts; one such advisory states that "persons in high risk categories (i.e., children, the elderly, and the immunocompromised) should not eat raw or lightly cooked sprouts" (52). In 1999 the FDA released two guidance documents for the sprouting industry describing methods for reducing microbial safety hazards associated with sprouts; these documents recommended the use of antimicrobial seed treatments and the testing of spent irrigation water for *Salmonella* spp. and *E. coli* O157:H7 (49). In 2000 a food safety training video intended for commercial sprout growers was released by the FDA in collaboration with the Food and Drug Branch of the California Department of Health Services (59). Despite these efforts, sprout-related outbreaks continue to occur in the United States on an annual basis. This has led to a recent plan to initiate rule making by the FDA in an attempt to reduce the number of sprout-related outbreaks of foodborne illness (55).

In this chapter, we discuss the native microflora found on sprout seeds and sprouts, the most promising antimicrobial interventions for sanitizing seeds and sprouts, and the detection technologies that have been studied. We also provide insights on the unique challenges and opportunities that seed sprouts afford from a food safety perspective and discuss strategies for reducing the risk of future sprout-related outbreaks of foodborne illness.

NATIVE MICROBES ASSOCIATED WITH SPROUTS

Through the use of traditional bacteriological methods, sprout seeds and seed sprouts are known to harbor high populations of a variety of indigenous microbes. Populations of total mesophilic aerobes on seeds destined for sprouting are often in the range of 4 to 5 $\log_{10}$ CFU/g, with coliforms comprising 10% or less of the population (2, 3, 134, 158). Fecal coliforms are often detected, with populations of up to 3 $\log_{10}$ CFU/g reported (124, 134). The fecal coliforms cultured from sprout seeds are primarily *Klebsiella pneumoniae, Enterobacter aerogenes,* and *Enterobacter agglomerans* (95, 124, 158). Native bacteria can be internalized within seeds. Mundt and Hinkle (115) found that 13% of surface-sterilized alfalfa seeds and 15% of surface-sterilized soybean seeds harbored internalized bacteria. A wide variety of molds also can contaminate the surfaces of sprout seeds (3).

Sprout surfaces are populated with higher numbers of cultivable native microflora than sprout seeds (see below) and harbor some of the highest populations of indigenous microflora reported for nonspoiled fresh produce at retail. Favorable temperatures and frequent watering cycles used in sprout propagation, as well as leaching of nutrients from seeds and roots during sprout development, favor rapid growth of native microbes. Increases of 1 $\log_{10}$ CFU/g for yeast and mold and 3 to 5 $\log_{10}$ CFU/g for aerobic bacteria 2 days after initiation of germination have been reported (3, 23). Similar to those isolated from other types of produce, native bacteria isolated from sprouts consist primarily of gram-negative rods and appear to originate mainly from the seed (104, 146, 158, 172). Yeast and mold are also present (Fig. 1). Additional sources of microbes include the air (124), irrigation water, and workers.

Populations of aerobic mesophilic bacteria on seed sprouts range from 5 to 12 $\log_{10}$ CFU/g but are usually between 7 and 9 $\log_{10}$ CFU/g (3, 8, 134, 158, 177). Populations of aerobic mesophilic bacteria determined for laboratory-grown and commercially grown sprouts at retail are similar (39, 64, 127). Populations of *Enterobacteriaceae* in the range of 7 to 8 $\log_{10}$ CFU/g have been reported (39, 155), and populations of total coliforms, fecal coliforms, and yeast and mold vary widely (4 to 8, <1 to 6, and 2 to 6 $\log_{10}$ CFU/g, respectively) (3, 8, 9, 39, 127, 134, 146, 155, 156, 158, 172).

Enterobacteriaceae and *Pseudomonas* spp. are the primary cultivable indigenous bacteria on a variety of sprout types (8, 9, 128). The predominant bacterial genera on sprouts during the germination process can change with time. Mølbak et al. (110) reported that *Erwinia* and *Paenibacillus* spp. were dominant on 1-day-old alfalfa sprouts but were gradually replaced by *Pseudomonas,* which predominated at days 3 and 5. In Sweden and the United States, the most common coliform isolated from sprouts is *Pantoea agglomerans* (27, 95). In addition to *P. agglomerans,* other fecal coliforms that have been routinely isolated from sprouts include *K. pneumoniae, E. aerogenes, Enterobacter cloacae, Enterobacter sakazakii,* and *Citrobacter freundii* (8, 124, 127, 142, 155, 158). As fecal coliforms are often found on plant surfaces, including those of sprout seeds and sprouts, their presence is not a good

Figure 1 Scanning electron micrograph of native yeast and rod-shaped bacteria on the surface of a mung bean sprout cotyledon.

indicator of recent fecal contamination (127, 193). Representatives of the bacterial genera *Achromobacter, Acinetobacter, Alcaligenes, Bacillus, Chromobacterium, Flavobacterium, Hafnia, Moraxella, Lactococcus, Lactobacillus, Leuconostoc, Paenibacillus, Rahnella, Ralstonia, Serratia, Sphingomonas,* and *Stenotrophomonas* have also been isolated from sprouts (8, 18, 104, 110, 120, 121, 127, 144, 177, 187). Lactic acid bacteria appear to normally comprise a minor component of the total microflora on sprouting seeds (8, 9, 127, 155).

There is a lack of studies using modern methods of microbial ecology to investigate the native microflora on sprout seeds and sprouts. Matos et al. (104) utilized community-level physiological profiling, a technique based on carbon source utilization patterns (70), to compare the native aerobic heterotrophic microbial communities associated with five sprout types. Based on this methodology, communities on alfalfa and clover sprouts are more similar to each other than to communities on sunflower, mung bean, and broccoli sprouts. Communities on the latter three sprout types are distinct from one another. Comparison of communities on alfalfa and clover sprouts grown from different seed lots or in different commercial facilities indicates that sprout type is responsible for more variability among communities than either seed lot or growing facility.

To our knowledge, cultivation-independent molecular techniques have not yet been applied for identifying microorganisms associated with seed sprouts. A recent analysis of GenBank entries indicated that only approximately half (27 of 53) of the discernible major phyla within the domain *Bacteria* have cultivated representatives (145). Plant phyllosphere and rhizosphere microbial communities are found to be more complex when examined by cultivation-independent methods than when examined by cultivation-based methods (91, 192). Thus, it is expected that the resident microflora on sprouts is more diverse than presently recognized. This idea was recently supported by results of two independent studies (105, 110) which compared direct microscope bacterial counts with agar plate counts of bacteria isolated from alfalfa sprouts. In most instances, only approximately 2 to 10% of the native bacteria present were cultivable.

Native bacteria are firmly attached to sprout surfaces, as evidenced by the inability of water washes after harvest to reduce populations by more than about 1 $\log_{10}$ CFU/g (8, 27, 126). Native microorganisms have been demonstrated by conventional scanning electron microscopy to be present both as solitary cells and as members of biofilms on sprout cotyledons, hypocotyls, and root surfaces (34, 43, 178) (Fig. 2). The native biofilms on sprout surfaces are heterogeneous assemblages of different bacteria (as determined based on various morphotypes), yeasts, or mixtures of bacteria and yeasts (38, 43). Using conventional scanning electron microscopy, Fett and Cooke (43) imaged native biofilms that were several cell layers thick. With the use of confocal scanning laser microscopy, native biofilms on alfalfa, clover, and mung bean sprouts were estimated to have a maximum thickness of 13 μm (45). Bacteria present in biofilms of even minimal thickness most likely have greatly increased resistance toward desiccation, predation, and washing and sanitizing treatments (30,111).

Certain indigenous bacteria also may gain protection from environmental and antimicrobial stresses by internalization into plants (67, 74). Entry into plants can be passive at sites such as wounds, at areas of secondary root emergence or broken trichomes, or at natural openings such as stomata and lenticels. Active entry by the action of bacterial hydrolytic enzymes such as pectinases and cellulases can also take place. Once inside of plants, the bacteria may reside in vascular tissue, in intercellular spaces, or in intracellular locations in a latent state or may actively colonize plant tissues (74). Gagne et al. (67) found that pseudomonads and *Erwinia*-like bacteria constitute the majority (75%) of the isolates obtained from the xylems of field-grown alfalfa, similar to the pattern of predominant bacteria associated with a

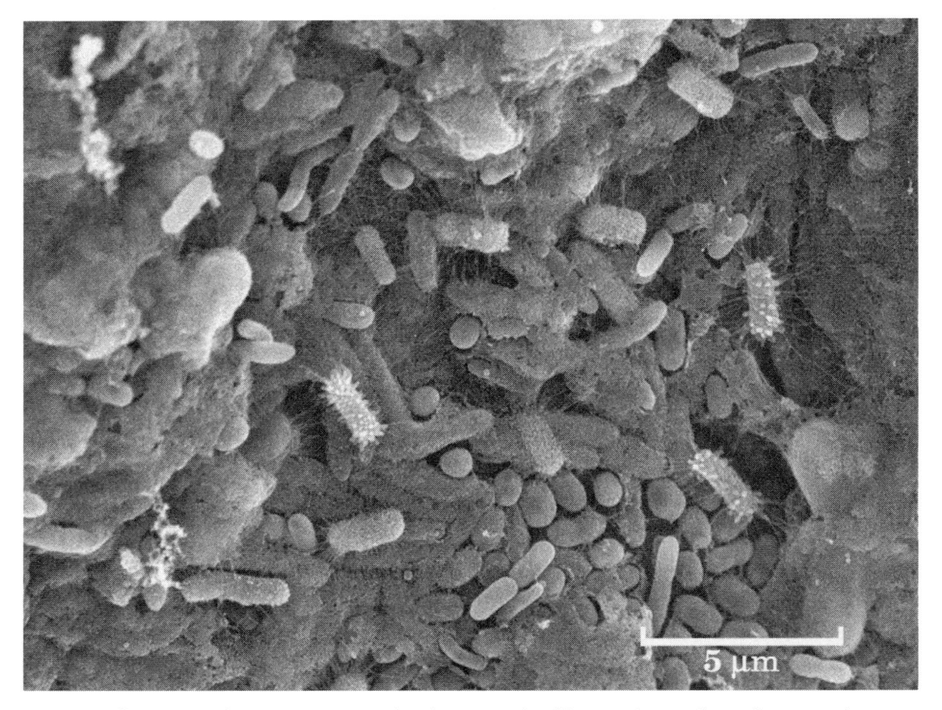

Figure 2 Scanning electron micrograph of a native biofilm on the surface of a mung bean sprout cotyledon.

variety of sprout types as noted above. Populations of bacteria isolated from the xylems ranged from 3.8 to 4.6 log_{10} CFU/g (67). Dong et al. (29) recently demonstrated that a strain of *K. pneumoniae* was a good colonizer of alfalfa roots and the interiors of the plants at 5 days into the sprouting process.

HUMAN PATHOGENS ASSOCIATED WITH SPROUTS

As noted above, consumption of contaminated raw seed sprouts has been responsible for numerous foodborne outbreaks of infection both in the United States and in other countries (for detailed listings of sprout-related outbreaks, see references 42, 103, 117, 148, 154, and 168). Sprout-related outbreaks have received much attention from regulatory bodies around the world since 1995. From 1995 through 2004, the number of annual outbreaks in the United States varied from a high of six to a low of one. The numbers of culture-confirmed cases of human illness per outbreak in the United States have ranged from 4 to almost 500. Recorded sprout-related outbreaks have also taken place in Canada, Denmark, Finland, Japan, The Netherlands,

Sweden, and the United Kingdom. Even though the great majority of sprout-related illnesses have been due to contamination with various serovars of *Salmonella,* the largest outbreak worldwide took place in Japan in 1996, with over 7,000 confirmed cases, and this outbreak was due to consumption of radish sprouts contaminated with *E. coli* O157:H7 (109, 181). The radish sprout industry in Japan has still not recovered from this devastating outbreak. Single outbreaks in the United States due to contamination of sprouts with *Bacillus cereus* and *Yersinia enterocolitica* have been recorded (131, 154). Enterotoxin-producing strains of *B. cereus* were recently isolated from 12 of 17 samples of soybean sprouts purchased in retail markets in Korea, but no associated outbreak was reported (93). In almost all outbreaks, contaminated seeds are thought to be the primary source of the bacterial pathogens based on direct isolation from the seeds and/or epidemiological evidence (117, 168). Seeds sold for sprouting are harvested from fields in numerous countries, including Australia, Burma, Canada, China, Italy, Peru, Thailand, and the United States.

Although not associated with sprout-related outbreaks, a number of other human pathogens have been isolated from sprouts. These include *Listeria monocytogenes* (163), *Aeromonas caviae* and *Aeromonas hydrophila* (19, 108), *Staphylococcus aureus* (134), *K. pneumoniae* (127, 142), and the protozoan parasites *Giardia* and *Cryptosporidium* spp. (140). In the United States, where regulatory agencies presently have a zero tolerance policy for *L. monocytogenes* on ready-to-eat foods (60), packages of sprouts have been recalled due to their potential contamination with this pathogen (53).

Several studies have demonstrated that *Salmonella* spp., *E. coli* O157:H7, and *L. monocytogenes* have the potential to grow to high levels on laboratory-grown sprouts propagated from artificially inoculated seeds (3, 5, 22, 23, 82, 90, 160, 161, 167, 179). Increases of up to 100,000-fold in populations of these microbes on sprouts propagated at 20 to 30°C have been noted, with the majority of growth taking place during the first 48 h of propagation. The doubling time for *S. enterica* on sprouting alfalfa seeds is estimated to be 47 min during the initial rapid growth phase. Growth is not dependent on serovar, isolation source, or virulence characteristics (82). Maximum populations for *S. enterica* and *E. coli* O157:H7 ranging from 5 to 8 $\log_{10}$ CFU/g have been reported. Populations of *E. coli* O157:H7 on alfalfa sprouts were consistently 1 log lower than those of *S. enterica.* This finding was attributed to stronger attachment of *Salmonella* to sprout surfaces, which lessened the rinsing effect of frequent irrigation (5, 25).

For *L. monocytogenes,* maximum populations on sprouting alfalfa seeds range from 5 to 8 $\log_{10}$ CFU/g (73, 120, 146). Gorski et al. (73) documented a wide range of abilities among strains of *L. monocytogenes* to attach to alfalfa sprout surfaces. These strain-specific differences were not related to serotype, lineage, or the original source of the pathogen (e.g., plant versus nonplant). Populations of *Salmonella*, *E. coli* O157:H7, and *L. monocytogenes* have been shown to be stable or to decline only slightly on contaminated sprouts stored at refrigeration temperatures (23, 82, 90, 146, 167).

The maximum populations achieved by bacterial human pathogens on sprouts do not appear to depend on the initial populations on the seeds (23, 73, 160) when the seeds are inoculated at a relatively high level. The findings indicate that there may be a maximum carrying capacity for pathogens on sprout surfaces. This may be due to the number of available attachment sites and the level of nutrients, both of which may be influenced by the nature and population of indigenous microbes. However, for seeds inoculated with lower levels, the extent of pathogen proliferation seems to be influenced by the initial inoculum dose. For example, Charkowski et al. (25) showed that, at 20°C, the level of *S. enterica* serovar Newport reached 3 $\log_{10}$ CFU/sprout on sprouts grown from seeds inoculated with 1 CFU of the pathogen/ml. On seeds inoculated with 2 $\log_{10}$ CFU/ml or higher levels, the final pathogen populations reached more than 5 $\log_{10}$ CFU/sprout.

Populations of *S. enterica* (1 to 4 $\log_{10}$ CFU/g) attained after sprouting naturally contaminated seeds by using growth conditions mimicking actual industrial propagation practices were several log units lower than those reported after using artificially inoculated seeds (1, 161). The lower final populations may be due to (i) a much lower initial pathogen level on the naturally contaminated seeds than on the artificially inoculated seeds used for such studies, (ii) a higher percentage of injured bacterial cells, and (iii) the use of different methods and increased frequency of irrigation that have a greater probability of rinsing off pathogens that were not tightly bound to the sprout surface.

Other human pathogens or opportunistic human pathogens shown to be able to multiply during growth of sprouting seed in the laboratory include *B. cereus* (attained populations of 4 $\log_{10}$ CFU/g on alfalfa and mung bean sprouts and 7.3 $\log_{10}$ CFU/g on rice sprouts [79, 128]), *K. pneumoniae* (attained populations of 5 to 6 $\log_{10}$ CFU/g on alfalfa and mung bean sprouts [124]), and *Vibrio cholerae* (attained a population of 6 $\log_{10}$ CFU/g on alfalfa sprouts [23]). Although occasionally isolated from raw sprouts, enterotoxigenic *Staphylococcus aureus* is not believed to be a food safety concern as the pathogen is unable to grow to high populations and produce enterotoxin on sprout surfaces due to competition with native microbes (117, 172).

S. enterica, *E. coli* O157:H7, and *L. monocytogenes* appear to preferentially colonize the roots of sprouting alfalfa seeds (25, 73). By the second day of sprouting, aggregates of *S. enterica*, but not *E. coli* O157:H7, were visualized on alfalfa roots (24). These aggregates may or may not represent actual biofilms consisting of bacterial cells attached to one another and to the plant surface by extracellular bacterial polymeric materials.

As for several other types of produce, including lettuce and tomatoes (180), pathogens can be internalized in sprouts during plant growth presumably via uptake through the root system. Itoh et al. (89) were the first to demonstrate this phenomenon for sprouts. By using immunofluorescence and scanning immunoelectron microscopy, *E. coli* O157:H7 was shown to locate in stomata and the vascular systems of radish sprouts grown from artificially inoculated seed. This was later confirmed for *Salmonella* spp. in mung bean and alfalfa sprouts by using bioluminescent and autofluorescent transconjugant strains inoculated onto seeds or roots (28, 69, 179). As neither *Salmonella* spp. nor *E. coli* O157:H7 is known to produce plant cell wall-degrading enzymes (e.g., cellulases and pectinases), entry into sprout roots is most likely due to passive uptake at sites of injury where lateral roots emerge (29, 74).

METHODS OF DECONTAMINATION OF SPROUT SEEDS

There have been more reported studies on interventions for the elimination of bacterial human pathogens from sprout seeds and sprouts than on those from any other type of produce. Since seeds are thought to be the primary source of pathogens for sprout-related foodborne outbreaks, more potential interventions have been tested on seeds than on sprouts. Adequate sanitation of seeds presents a unique challenge in the produce area. Due to the ability of surviving bacteria to grow rapidly during the sprouting process, the goal is to completely eliminate any pathogens present on the seeds. This must be accomplished while maintaining posttreatment seed viability as well as a commercially acceptable sprout yield, appearance, and shelf life. Taking into account the nature of the commercial sprouting industry (comprising primarily small firms with 10 employees or fewer [170]) as well as the fact that some sprouts are grown by consumers at home, seed interventions should be inexpensive, present no hazard to the environment or human health, be easily applied, and preferably consist of a single step. The ideal intervention would be effective against bacterial human pathogens and parasites that have been associated with sprouts as well as viruses that have been implicated in other produce-related outbreaks (e.g., hepatitis A) but not yet associated with sprouts. For organic sprout growers, restrictions on chemical use by private organic certifying organizations and certification under the U.S. Department of Agriculture Organic Rule also need to be addressed. More expensive interventions in terms of equipment and material costs may be feasible for application by large seed distributors before shipment of seeds to customers. In this section, we will discuss studies on antibacterial interventions for seeds destined for sprouting and point out interventions that appear to have the most potential for use by U.S. sprout growers. The reader is also referred to the recent review on this subject by Fett (42).

Chemical and Physical Interventions

A wide variety of chemical and physical interventions have been tested as stand-alone treatments for eliminating *Salmonella, E. coli* O157:H7, and *L. monocytogenes* from artificially inoculated sprout seeds. Numerous aqueous chemicals have been tested as seed-sanitizing solutions. The FDA guidance documents (49) recommend that growers treat sprouting seeds with an approved antimicrobial treatment such as 20,000 ppm of free chlorine from $Ca(OCl)_2$ immediately before sprouting. The use of chlorine at up to 20,000 ppm and that of gamma irradiation at up to 8 kGy (50) are two approved methods for sanitizing seeds before sprouting for human consumption. The use of 20,000 ppm of free chlorine is the standard against which the efficacy of other aqueous sanitizers has been compared.

Chemical interventions

In addition to chlorine [NaOCl and $Ca(OCl)_2$], natural and synthetic antimicrobials that have been tested as aqueous sanitizers for alfalfa seeds include acetic acid, acidic electrolyzed water, acidified ClO_2, acidified $NaClO_2$, $Ca(OH)_2$, calcinated calcium, Califresh-S, ClO_2, Citrex, citric acid, Citricidal, Citrobio, Environné, ethanol, Fit, H_2O_2, H_2SO_4, lactic acid, Na_2CO_3, ozonated water, thyme oil, trisodium phosphate, Tsunami, Tween 80, Vegi-Clean, and Vortexx (6, 10, 11, 12, 44, 69, 81, 92, 95, 122, 147, 149, 150, 151, 153, 159, 166, 178, 185). Supercritical CO_2 technology has also been evaluated (107). A number of aqueous chemicals have been reported to be as effective as 20,000 ppm of chlorine for reducing populations of bacterial human pathogens on artificially inoculated alfalfa seeds without significantly reducing the seed germination rate. These include 1% $Ca(OH)_2$, 1% calcinated calcium, Citrex (20,000 ppm), 8% H_2O_2, and Fit (12, 44, 81, 147, 166, 185) (Table 1). Pretreatment of seeds with surfactants or the addition of surfactants to sanitizer solutions either had no effect or improved efficacy only slightly (1 log unit or less) (40, 166, 185). The use of sanitizers at an elevated temperature (55°C) with or without sonication led to increased efficacy but also decreased the alfalfa seed germination rate (11). Presoaking of seeds in water for 1 h before treatment with 20,000 ppm of chlorine increased the killing by 2 log units but also led to a significant decrease in the seed germination rate (40). Treatment of alfalfa seeds inoculated with *E. coli*

Table 1 Comparison of selected aqueous chemical treatments with $Ca(OCl)_2$ for sanitizing alfalfa seed

Sanitizer	Concn	Treatment time (min)	Pathogen	Log_{10} reduction	Reference
H_2O_2	8%	10	*Salmonella*	3.3	81
$Ca(OH)_2$	1%	10	*Salmonella*	3.8	81
$Ca(OCl)_2$	20,000 ppm[a]	10	*Salmonella*	3.9	81
H_2O_2	8%	10	*E. coli* O157:H7	2.9	81
$Ca(OH)_2$	1%	10	*E. coli* O157:H7	3.2	81
$Ca(OCl)_2$	20,000 ppm[a]	10	*E. coli* O157:H7	2.5	81
Calcinated calcium	1%	10	*Salmonella*	2.9	185
$Ca(OCl)_2$	20,000 ppm[a]	10	*Salmonella*	2.0	185
Fit	Per manufacturer	15	*Salmonella*	2.3	12
$Ca(OCl)_2$	20,000 ppm[a]	15	*Salmonella*	2.3	12
Fit	Per manufacturer	15	*E. coli* O157:H7	>5.4	12
$Ca(OCl)_2$	20,000 ppm[a]	15	*E. coli* O157:H7	2.6	12
Citrex	20,000 ppm	10	*Salmonella*	3.6	44
$Ca(OCl)_2$	16,000 ppm[b]	10	*Salmonella*	3.4	44
Citrex	20,000 ppm	10	*E. coli* O157:H7	3.4	44
$Ca(OCl)_2$	16,000 ppm[b]	10	*E. coli* O157:H7	3.3	44

[a]Prepared in 50 mM potassium phosphate buffer, pH 7.0.
[b]Prepared in 500 mM potassium phosphate buffer, pH 6.8.

O157:H7 with a bacteriocin (colicin HU195, an E2 type) produced by a generic strain of *E. coli* at 10,000 AU/g gave a reduction in levels of one strain of greater than 5 log_{10} units, but the bacteriocin was much less effective against two other strains inoculated onto alfalfa seeds (116). The results underscore the issue of differential strain susceptibility to bacteriocins that might restrict the efficacy of bacteriocins as antimicrobials under commercial conditions.

There have been few studies on the effects of aqueous chemical sanitizers as stand-alone treatments for other sprout seed types. Fett (41) found that treatment of artificially inoculated mung bean seeds with high levels of aqueous chlorine (16,000 to 18,000 ppm) led to log reductions in levels of *E. coli* O157:H7 and *Salmonella* (4 to 5 log_{10} CFU/g) that were higher than those demonstrated for alfalfa seeds in similar previous studies in the same laboratory. Bari et al. (6) reported that treatment of mung bean seeds with acidic electrolyzed water in combination with sonication led to reductions of 4 log_{10} units. A similar treatment of radish and alfalfa seeds led to reductions of only 1.5 and 2.6 log_{10} units, respectively.

Results of combination treatments with aqueous sanitizers have rarely been reported. A sequential treatment of inoculated alfalfa seeds with aqueous

ClO_2 (25 mg/liter; 5 min), ozonated water (14.3 mg/liter; 3 min), and thyme oil suspension (5 ml/liter; 3 min) led to a 3.5- to 4-$\log_{10}$-unit reduction in levels of *E. coli* O157:H7 and was much more effective than each of the chemicals used alone (153). However, after 3 days of sprouting of treated seeds, populations of the pathogen were almost 8 $\log_{10}$ CFU/g. Unless able to eliminate pathogens, such combination treatments may not be practical for commercial growers due to the extra costs and manipulations involved.

The comparison of efficacy data for various aqueous sanitizers is valid for studies within a laboratory using a single set of experimental protocols, but comparison of efficacy data for sanitizers tested in different laboratories may not be valid. This is due to differences in methodologies, including differences in the seed lots employed (lots with more damaged and wrinkled seeds are more difficult to sanitize [24]) (12), in the methods of seed inoculation, in the drying and storage times before treatments are applied, in the initial pathogen populations on the seeds, in the organic loads on the seeds (11), in the use of rinses before and after seed treatment, in the pHs of sanitizer solutions, in the seed-to-sanitizer volume ratios, in the methods for applying the sanitizer treatments (e.g., hand mixing versus mechanical mixing), and in media used to enumerate survivors. In addition, if injured cells are not effectively resuscitated and accounted for, the efficacy of the intervention might be overestimated.

As evidence of the degree of variability in the results obtained in different laboratories, reported reductions in levels of *Salmonella* and *E. coli* O157:H7 artificially inoculated onto alfalfa seeds treated with 16,000 to 20,000 ppm of free chlorine range from approximately 2 to 7 $\log_{10}$ CFU/g (40, 44, 69, 92, 95, 166, 185). Log reductions for the two pathogens are usually similar (1-$\log_{10}$-unit difference or less) for the same seed treatments applied in the same laboratory. The lower-log reductions reported were most likely due, at least in part, to the low inoculum levels (2 to 3 $\log_{10}$ CFU/g) before treatment as the maximum population reductions that could be demonstrated under these conditions were 2 to 3 $\log_{10}$ units. The higher values may be due in part to inadequate rinsing or neutralization of residual chlorine after seed treatment. Some researchers have utilized chlorine solutions adjusted to a neutral pH and some have not. At a pH of 6.8 and a temperature of 20°C, approximately 75% of chlorine in solution is in the form of hypochlorous acid, with the other 25% present as hypochlorite ion, a less effective antibacterial form of chlorine (31). At more basic pHs, the percentage of the less effective hypochlorite ion increases while the percentage of hypochlorous acid decreases.

The efficacy of treatments is also dependent on the type of seed. Higher-log reductions in levels of *Salmonella* and *E. coli* O157:H7 were observed after chlorine treatment of inoculated mung bean seeds than after treatment of alfalfa seeds (40, 41). In a recent in-depth analysis of the published literature concerning sprout seed sanitization, Montville and Schaffner (112) concluded that chemical treatment efficacy data are highly variable and the data are more variable when more published information is available. With the exception of chlorine, most chemical treatments have been examined by only a single laboratory. The one consistent finding among the different laboratories is that no single aqueous chemical treatment has proven to be capable of eliminating bacterial human pathogens from sprouting seeds without significant reductions in seed germination rates.

The use of naturally contaminated seeds for sanitization studies is preferable to that of artificially contaminated seeds, but naturally contaminated sprout seeds are usually not readily available. Thus, there have been few studies on the effectiveness of aqueous chemical sanitizers at eliminating pathogens from naturally contaminated seeds, and the results to date have been conflicting. Both Suslow et al. (164) and Fett (40) reported that treatment of alfalfa seeds naturally contaminated with *S. enterica* with 2,000 or 20,000 ppm of chlorine from $Ca(OCl)_2$ eliminated the pathogen. A similar study by Stewart et al. (161) using an identical contaminated seed lot indicated that treatment with 20,000 ppm of chlorine did not eliminate *Salmonella*. There are several possible explanations for the conflicting results, including differences in the extents of mixing during treatment, differences in the extents of contamination of the seed samples tested, the location of the pathogens on individual contaminated seeds, and the distribution of the pathogen throughout the lot.

Based on an investigation of a 1999 outbreak of salmonellosis related to contaminated clover sprouts, the use of 20,000 ppm of chlorine for seed soaking as recommended by the FDA (49) and as applied by commercial growers may not be totally effective at eliminating the risk of sprout-related outbreaks of foodborne illness (133). However, an investigation of a 1999 alfalfa sprout-related outbreak of salmonellosis indicated that sprouts produced by growers who used a chlorine seed soak were not involved in the outbreak while those produced by growers who did not use a seed decontamination step were (71). It is difficult to assess the efficacy of the chlorine treatment under commercial practice due to variability in the actual preparations and methods of application of the chlorine solutions.

Gas and vapor treatments for sanitizing seeds have also been examined, and one would expect that such treatments would have improved capabilities of reaching pathogens present deep in cracks in the seed coat and in natural openings such as the hilum. Chemicals tested include gaseous acetic acid, allyl isothiocyanate, ammonia, carvacrol, cinnamic aldehyde, eugenol, linalool, methyl jasmonate, thymol, and *trans*-anethole (28, 80, 125, 186).

Two promising treatments are gaseous acetic acid for sanitizing mung bean seeds and gaseous ammonia for sanitizing alfalfa and mung bean seeds. Treatment of mung bean seeds with gaseous acetic acid (242 μl per liter of air) for 12 h at 45°C eliminated both *Salmonella* and *E. coli* O157:H7 (initial populations of 3.7 to 6.0 $\log_{10}$ CFU/g) and resulted in a 4-$\log_{10}$-unit reduction in levels of *L. monocytogenes* from artificially inoculated seeds without a reduction in seed germination rates (28). A similar treatment of alfalfa seeds led to a significant reduction in the seed germination rate (P. Delaquis, personal communication). Treatment of inoculated alfalfa and mung bean seeds with 300 ppm of gaseous ammonia (22 h at 20°C) led to a 2- to 3-$\log_{10}$-unit reduction in levels of *Salmonella* and *E. coli* O157:H7 on alfalfa seeds and a 5- to 6-$\log_{10}$-unit reduction in levels on mung bean seeds without reducing seed germination rates (80).

Physical interventions

A variety of physical treatments have been tested as stand-alone interventions for sanitizing sprout seeds, including hot water, dry heat, gamma irradiation, hydrostatic pressure, pulsed UV light, radio frequency dielectric heating, and ultrasound (sonication). Heat treatments of planting materials, including seeds, prior to planting for the purpose of eliminating plant pathogens (sometimes referred to as thermotherapy) have been studied since at least the 1920s (106). Hot-water treatments for reducing populations of bacterial human pathogens on sprout seeds were first examined by Jaquette et al. (90). They found that a 5-min treatment in hot water at 57 to 60°C led to a 2.5-$\log_{10}$-unit decrease in the population of *S. enterica* on artificially inoculated alfalfa seeds and no appreciable reduction in the seed germination rate. However, slightly higher temperatures or longer treatment times led to significant reductions in the germination rate. Enomoto et al. (33) studied the ability of hot-water treatments to eliminate the nonpathogenic *E. coli* strain ATCC 25922 from artificially inoculated alfalfa seeds. By using a three-step process (30-min presoak at 25°C, 9-s treatment at 50°C, and 9-s treatment at 85°C), a >4-$\log_{10}$-unit reduction was achieved with no significant loss in germination rate. This reduction was greater than that obtained in the same study using 20,000 ppm of chlorine. Mung bean seeds artificially inoculated with *S. enterica* serovar Senftenberg 775W, an unusually heat-resistant strain of *Salmonella*, were also successfully treated with hot water (184). Decimal reduction times (the time of treatment to obtain a decrease of 90% in numbers of viable CFU) of 3.9, 1.9, and 0.6 min were determined for treatments at 55, 58, and 60°C, respectively. Reductions of >5 $\log_{10}$ units without reduction of the seed germination rate were obtained with the following temperature and time regimes: 55°C and 20 min; 60°C and 10 min;

70°C and 5 min; and 80°C and 2 min. Excellent results have also recently been reported by Hu et al. (83) for dry-heat treatments of mung bean seeds. Storage of seeds inoculated with cocktails of strains of *E. coli* O157:H7 or *S. enterica* serovars at 55°C for 4 or 5 days, respectively, completely eliminated initial pathogen populations of 6 $\log_{10}$ CFU/g for *E. coli* O157 and 4 $\log_{10}$ CFU/g for *Salmonella.*

On the basis of these studies, hot-water and hot-air treatments of sprout seeds appear promising; however, the commercial application of such strategies may be problematic. Even though hot-air treatments were successful at eliminating pathogens from mung bean seeds, the authors reported in the same study that the germination rate of alfalfa seeds treated in the same manner was significantly reduced. Heat treatments may be more applicable to larger sprout seed types with thicker seed coats where the storage tissues and embryo are more insulated against the damaging effects of heat. In an in-depth series of studies on the application of hot humid air (aerated steam) to seeds to control seed-borne diseases of cereals, Forsberg (61) determined that there are differences in heat tolerance levels not only among seed types but also within a seed lot due to the production and storage histories of the seeds. Seeds of low moisture content are more heat resistant. Seeds in storage for longer time periods, even under ideal conditions (low temperature and low relative humidity), undergo an aging process which reduces their ability to withstand stress, including heat. Where final seed lots ready for distribution consist of mixtures of seeds from different primary lots (as is sometimes the case in the seed sprout industry), variations in heat tolerance levels within the final lot can occur. Thus, there is often a very narrow time-temperature treatment window where pathogen levels can be significantly reduced or eliminated from a particular seed lot and the seed germination rate, quality of the plant, and yield can be maintained at acceptable levels. Because of these considerations, Forsberg (61) recommended that each seed lot be sampled at various locations and that the samples of seeds be pretested for heat tolerance before the entire seed lot is treated. If heat treatments are applied by large seed distributors, a decrease in the subsequent storage potential of the seeds may result (106).

In addition to the difficulty in maintaining the seed germination rate and the yield at acceptable levels, the ability of hot-water treatments to eliminate *S. enterica* from naturally contaminated seeds has been questioned. Suslow et al. (164) found that placing naturally contaminated seeds in hot water at up to 85°C for 1 min did not eliminate *Salmonella.* Also, a recent sprout-related outbreak of salmonellosis was reportedly due to consumption of sprouts grown from contaminated alfalfa seeds that had been treated with a proprietary hot-water treatment followed by a soak in aqueous chlorine (2,000 ppm) (188).

The FDA permits doses of ionizing radiation up to a maximum absorbed dose of 8 kGy to control human pathogens on sprout seeds (50). Treatment of sprout seeds with gamma irradiation would most likely be applied before distribution of seeds to individual sprout growers. Treatment with gamma radiation at 2 kGy can significantly reduce (2 to 3 $\log_{10}$ units) bacterial pathogen populations on sprouting seeds while maintaining adequate sprout yield and nutritional quality (37, 138, 169). Exposure to higher levels of gamma radiation are required to achieve a 5-$\log_{10}$-unit reduction on artificially inoculated seeds and to eliminate *S. enterica* from naturally contaminated seeds, but the use of higher doses is not practical due to the harmful effects on yield and sprout quality. However, a recent study by Bari et al. (6) indicated that a dry-heat treatment (50°C; 1 h) followed by exposure to gamma radiation (2 to 2.5 kGy) eliminated *E. coli* O157:H7 from artificially inoculated alfalfa, radish, and mung bean seeds (4- to 5-$\log_{10}$-unit reductions) without decreasing seed germination rates. The combination treatment did significantly decrease the lengths of both radish and mung bean sprouts, but not that of alfalfa sprouts, after 4 days of growth. A problem with the commercial application of gamma radiation for sanitizing seeds is the phenomenon of uneven absorbance of dosages by seeds at different locations in the treatment chamber. The dose absorbed by seeds located near the exterior of the chamber is higher than the dose absorbed by seeds at the center of the chamber. This unevenness of exposure might lead to variable effects on pathogen reduction within the treated batch of seeds as well as differential effects on the subsequent seed germination rate, sprout quality, and yield.

High-hydrostatic-pressure treatments may be useful for sanitizing a limited number of types of sprout seeds before distribution. Exposure to high pressure (250 to 300 MPa for 15 min at 20°C) significantly reduced (>6 $\log_{10}$ units) populations of *S. enterica, E. coli,* and *Listeria innocua* on garden cress seeds, but the treatment was highly detrimental to the germination of radish, mustard, and sesame seeds (191). High-hydrostatic-pressure treatment of alfalfa seeds (575 MPa for 2 min or 475 MPa for 2 to 8 min at 40°C) was less effective at reducing *E. coli* O157:H7 and *L. monocytogenes* populations and led to a significant reduction in the germination rate (4).

Combined chemical and physical treatments

There have been few published studies on combining chemical and physical treatments for seed sanitization. Combining two or more antimicrobial treatments that have different modes of action may result in increases in killing that are either additive or synergistic. Synergistic effects occur if the first treatment leads to a large number of injured cells that have increased susceptibility to the second antimicrobial intervention. Bari et al. (6) reported that a combination treatment with dry heat (50°C; 1 h), hot acidic electrolyzed water, and sonication eliminated (a 4.6-$\log_{10}$-unit reduction) *E. coli* O157:H7 from inoculated mung bean seeds without a reduction in germination rate or average sprout length after 4 days of growth but was unable to eliminate the pathogen from inoculated radish and alfalfa seeds. The dry-heat treatment alone and a combined treatment with dry heat followed by hot acidic electrolyzed water were not as effective. Sharma et al. (150) reported that treatment of ozone-sparged alfalfa seeds with dry heat (60°C; 3 h) was more effective than ozone treatment alone and resulted in a 4- to 4.8-$\log_{10}$-unit reduction in the *E. coli* O157:H7 population without compromising the germination rate. Combining sonication with hot aqueous chemical treatments led to slight increases in effectiveness (147). The combined effects of ionizing radiation treatments applied before seed distribution followed by various chemical treatments that could be applied by growers immediately before sprouting of seeds should be evaluated.

Biological Interventions

There have been few studies on the use of biological control (commonly referred to as competitive exclusion or probiotics in the food microbiology literature) strategies for controlling the growth and survival of human pathogens on produce, including sprouting seeds. In contrast, biological control has been intensively studied as a means of controlling plant diseases since at least the early 1920s (20) and, more recently, has been investigated for controlling human pathogens in chicks, swine, calves, meat, and dairy products (13, 118, 173). For control of human pathogens in food animals or in foods, single microbial strains or defined or undefined consortia of microbes have been tested as antagonists. Several commercial products are on the market for use in controlling the colonization of newly hatched chicks with *Salmonella*. One such product is named PREEMPT and is a mixture of 29 different bacterial isolates originally obtained from the cecal contents and cecal tissues of adult chickens (118). Effective biological control products for controlling growth of bacterial human pathogens on sprouting seeds and sprouts postharvest would be very useful for organic as well as conventional growers.

As demonstrated by several studies mentioned previously, bacterial human pathogens are surprisingly good competitors on sprout surfaces and can grow to high levels on growing sprouts after germination of artificially inoculated or naturally contaminated seeds. However, competitive exclusion does occur naturally on sprouts as evidenced by little growth (1 $\log_{10}$ unit or less) of bacterial human pathogens inoculated onto sprouts 1 to 5 days into the sprouting process (23, 146). Thus, the key for effective control is ensuring the presence of high levels of effective competitive microflora on the seed at the time of sprouting.

Several lactic acid bacteria isolated from alfalfa seeds and sprouts were demonstrated to be highly inhibitory toward *S. enterica, E. coli* O157:H7, and *L. monocytogenes* in vitro (121, 187); however, when the inhibitory isolate *Lactobacillus lactis* SP 26 was applied to seeds artificially inoculated with *L. monocytogenes* and the seeds sprouted, outgrowth of the pathogen was reduced by only 1 $\log_{10}$ unit (120). As mentioned above, lactic acid bacteria usually do not constitute a significant proportion of the native bacteria present on sprouting seeds and may not be able to outcompete or exhibit significant antibiosis against bacterial human pathogens on the growing plant surface. There is, however, a commercial product based on lactic acid bacteria that is sold in Japan for controlling bacterial human pathogens on sprouting seeds.

The proof that competitive exclusion is a viable intervention against *Salmonella* outgrowth on sprouts comes from the work of Matos and Garland (105). Addition of *Pseudomonas fluorescens* 2-79 to the seed soaking solution at 8 $\log_{10}$ CFU/ml led to a 4-$\log_{10}$-unit reduction in the population of *Salmonella* at days 1 and 3 of sprouting of artificially inoculated alfalfa seeds. Whole microbial communities isolated from alfalfa sprouts purchased at a supermarket and added to the seed soaking solution were not as effective at inhibiting outgrowth of *Salmonella* at days 1 and 3 but gave greater log reductions at day 7 of sprouting. There was no adverse effect of inoculation with strain 2-79 on sprout appearance. Strain 2-79 was originally isolated from the rhizosphere of a field-grown wheat plant and is a known biological control agent against the wheat root pathogen *Gaeumannomyces graminis* var. *tritici* (171). Recent studies by Fett (W. F. Fett, unpublished results) using a collection of mutants of strain 2-79 confirmed antibiosis of strain 2-79 against *S. enterica* both in vitro and in situ on sprouting alfalfa seeds and also indicated that the production of iron-binding siderophores and the antibiotic phenazine-1-carboxylic acid (the two primary antifungal metabolites produced by this strain) is not responsible for inhibitory activity. Such biocontrol strategies may also be effective for controlling bacterium-mediated spoilage of germinating seeds (32) and, with the use of endophytic antagonists, may be effective against internalized bacterial human pathogens (26).

Another biological approach to controlling pathogen survival and growth on sprouting seeds is the application of bacteriophage. The use of lytic phage for the control of bacterial plant pathogens is an active area of research (72). The application of lytic phage is also being studied for the control of bacterial human pathogens on fresh-cut produce. Spraying a lytic phage (8 $\log_{10}$ CFU/ml) onto inoculated fresh-cut honeydew melon was demonstrated to reduce populations of *L. monocytogenes* to below detectable levels, and no significant outgrowth of any survivors occurred during subsequent storage for 7 days at 10°C (97). Pao et al. (123) tested the abilities of two lytic phages to inhibit outgrowth of *S. enterica* on sprouting broccoli and radish seeds. Application of phage at approximately 6 to 7 $\log_{10}$ PFU/ml at the time of germination was not highly effective (less than a 1.5-$\log_{10}$-unit reduction after 24 h) against susceptible *S. enterica* serovars, and there was no activity against resistant *S. enterica* serovars. The host specificity of lytic phage would necessitate the use of phage mixtures under commercial practice.

A general observation concerning most antibacterial intervention studies published to date is that the effects of antibacterial treatments on seed germination rates only are reported. From a grower's perspective, the appearance, yield, and shelf life are also important parameters to consider.

INTERVENTIONS DURING SPROUTING OR POSTHARVEST

During sprouting, growing plants are frequently irrigated to prevent desiccation and to cool sprouts grown in large bins (primarily mung bean). The addition of antimicrobials to the irrigation water is a possible means of reducing the growth of bacterial human pathogens; however, there have been few published studies in this area compared to the number of studies on potential interventions for seeds. Spray irrigation of sprouting rice with chlorinated water (100 ppm) every 6 h was not effective at reducing populations of *B. cereus* or *L. innocua* (129). Daily spray irrigation with chlorine (100 ppm) reduced populations of *S. enterica* and *V. cholerae* by 2 $\log_{10}$ units or less on sprouting alfalfa seeds (22, 69). Taormina and Beuchat (167) tested the addition of several antibacterial chemicals to the spray irrigation water, including NaOCl (up to 2,000 ppm), $Ca(OCl)_2$ (up to 2,000 ppm), and acidified $NaClO_2$ (up to 1,200 ppm). None of the chemicals tested significantly reduced populations of *E. coli* O157:H7 on sprouting alfalfa seeds. When sprouting alfalfa seeds were rinsed with aqueous ClO_2 (25 ppm) or ozonated water (9 ppm) after 48 or 72 h of sprouting, populations of *E. coli* O157:H7 were not reduced (153). Rinsing with thyme oil (5 ppm) alone or in sequence with ClO_2 and ozonated water at 24 to 48 h into the sprouting process led to reductions in pathogen populations on the sprouts of up to 2 $\log_{10}$ units, but the same treatments were ineffective when applied to sprouts at 72 h (153). This may indicate that the pathogen had either formed biofilms or become part of biofilms formed by native microflora (38, 43, 45) and/or was internalized and thus resistant or not exposed to the sanitizing rinses. The effects of such a sequential treatment on sensory qualities of treated alfalfa sprouts were not evaluated.

The addition of antimicrobials to the irrigation water would not be compatible with the FDA recommendation to test spent irrigation water for

Salmonella and *E. coli* O157:H7 (49). Pathogens rinsed from the sprout surface and thus suspended in an aqueous solution would be more easily killed than would pathogenic cells still tightly attached to the sprout surfaces. Thus, a negative test result for the spent irrigation water would not accurately reflect the microbial safety status of the sprouts.

The reduction or elimination of bacterial human pathogens present on sprouts postharvest has also proven problematic. Few antimicrobial interventions have been reported to be capable of achieving a 5-$\log_{10}$-unit reduction or greater without an adverse impact on quality. Rinsing with water reduces bacterial human pathogens and native microbes on sprouts by 1 $\log_{10}$ unit or less (5, 126). Washing of alfalfa sprouts with acidic electrolyzed water (50 ppm of available chlorine) for 64 min resulted in a 3-$\log_{10}$-unit reduction in the *E. coli* O157:H7 population with no reported change in sprout appearance (149). Treatment with acidic electrolyzed water (84 ppm of available chlorine; 10 min) in combination with sonication and removal of seed coats led to a 3.3-$\log_{10}$-CFU/g reduction in *Salmonella* populations compared to the control (92). For *S. enterica* and *L. monocytogenes* on mung bean sprouts, a 10-min treatment with chlorous acid (268 ppm) resulted in an approximate 5-$\log_{10}$-unit reduction without adversely affecting visual quality (96).

For pathogens that are internalized, treatment with gamma radiation may be the only effective intervention. Rajkowski and Thayer (136) demonstrated that gamma irradiation to a minimum dose of 0.5 kGy could eliminate *S. enterica* from alfalfa sprouts grown from naturally contaminated seeds. Subsequently, exposure of uninoculated alfalfa sprouts to 2 kGy of gamma radiation was reported to extend the shelf life by more than 10 days without a reduction in nutritional quality (36, 137). Schoeller et al. (146) found that irradiation of alfalfa sprouts inoculated with *L. monocytogenes* (6 $\log_{10}$ CFU/g) with 3.3 kGy of beta radiation (electron beam) eliminated the pathogen without an adverse effect on quality. *Salmonella* and *E. coli* O157:H7 were eliminated from inoculated mung bean and radish sprouts with doses of 1.5 and 2.0 kGy, respectively, and acceptable quality was maintained (7). However, the use of gamma or beta radiation for eliminating bacterial human pathogens from sprouts has not yet received regulatory approval in the United States.

Few synthetic chemicals are allowable for use on organic produce under the USDA Agricultural Marketing Service's National Organic Program Regulations (176). These include ethanol and chlorine materials such as calcium and sodium hypochlorite and chlorine dioxide, hydrogen peroxide, and peracetic acid (for disinfecting equipment, seeds, and asexually propagated materials). Residual chlorine levels in the wastewater should not exceed the maximum residual disinfectant limit under the Safe Drinking Water Act, presently 4 ppm (34). Treatment with ionizing radiation is not allowed.

PATHOGEN SAMPLING AND DETECTION

Testing of Sprouts and Spent Irrigation Water for the Presence of Pathogens

Since the recommended use of 20,000 ppm of $Ca(OCl)_2$ for sanitizing seeds does not guarantee the elimination of pathogens (95, 161, 166), a microbiological testing program is also recommended to ensure sprout safety (49, 117). The presence of fecal coliforms is not a good indicator of recent fecal contamination because the coliforms are often found on plant surfaces as part of the native microflora (127, 193); therefore, the pathogens need to be targeted. In 1999, the FDA issued guidance to the sprout industry, recommending the testing of each production batch for the presence of *Salmonella* and *E. coli* O157:H7, the two most common agents of sprout-associated illness (49).

The optimal time for testing sprouts is when the pathogen level is highest but also when it is early enough to ensure that results are obtained before the product is shipped. The sprouting process encourages microbial proliferation and provides a natural environment for amplifying pathogens. Typically, the greatest increase in pathogen populations occurs during the first 24 h of sprouting, and maximal levels are achieved at 48 h (23, 25, 65, 68, 82, 90, 160, 161, 167). Table 2 summarizes published data regarding pathogen levels during sprouting at 24 h versus 48 h under various experimental growing conditions. In most cases, the levels of pathogens observed at 24 h closely approached those observed at 48 h. However, the differences between the levels of *Salmonella* at 24 and 48 h seemed to be greater (up to 2.3 $\log_{10}$ units) when naturally contaminated seeds were used than when artificially inoculated seeds were used. The FDA recommends sampling for microbiological testing 48 h after starting production (49).

Sampling of sprouts during seed germination in commercial settings presents certain difficulties. The distribution of pathogens in food is often nonhomogeneous; therefore, it is necessary to obtain multiple samples at various sites to ensure good representation of the entire production batch. Taking samples of sprouts may be difficult because the equipment used (e.g., rotary drums) is not designed for easy access and, in some cases, removal of sprouts from the production batch at various locations may disrupt proper seed germination (J. Louie, personal communication). Furthermore, additional equipment is required for preparing the samples for analysis (e.g., a homogenizer or stomacher). On the other hand, during production, the sprouts are

Table 2 Studies on the growth of bacterial human pathogens on sprouting seeds

Sprouting apparatus	Seed status (lot)	Pathogen	Log CFU/g of seeds (day 0)	Log CFU/g of sprouts		Reference
				Day 1 (24 h)	Day 2 (48 h)	
Mini drum, (25°C)	Naturally contaminated (COA98)	*S. enterica* serovar Muenchen	−1.5	2.0	2.0	65
Mini drum (30°C)	Naturally contaminated (45197)	*S. enterica* serovar Mbandaka	−2.4 (run 1), −1.5 (run 2)	−0.6 (run 1), 0.5 (run 2)	−1 (run 1), 1.8 (run 2)	65
Glass jars	Naturally contaminated (COA98)	*S. enterica* serovar Muenchen	−1	1.4	2.3	65
Glass jars	Naturally contaminated (45197)	*S. enterica* serovar Mbandaka	−1.6	2.6	3.1	65
Glass jars	Naturally contaminated (A)	*Salmonella*	−1.2	1	2.5	161
Glass jars	Naturally contaminated (B)	*Salmonella*	Below detection limit	1.7 to 2.2	2 to 4.5	161
Plastic container	Artificially inoculated	*S. enterica* serovar Stanley	3.2	6	7	10
Stainless-steel tray	Artificially inoculated	*S. enterica* serovar Typhimurium	3.4	5	5.4	23
Test tube	Artificially inoculated	*S. enterica* serovar Newport	1.8	4.7	6.3	25
Easy Green sprouting unit	Artificially inoculated	*S. enterica* serovar Cubana	0	5.7	5.7	82
Screen tray	Artificially inoculated; treated with 20,000 ppm of chlorine	*S. enterica* serovar Stanley	3.6	6.9	7.9	68

Plastic box with drain holes	Artificially inoculated; sprayed with water	*E. coli* O157:H7	3.1	5.0	4.8	167
Stainless-steel tray	Artificially inoculated	*E. coli* O157:H7	3.1	5.7	5.7	23
Glass jars	Artificially inoculated	*E. coli* O157:H7	3.9 (high inoculum); 1.9 (low inoculum)	6 (high inoculum); 4.5 (low inoculum)	5.5 (high inoculum); 5.3 (low inoculum)	160
Test tube	Artificially inoculated	*E. coli* O157:H7	2.7	3.8	4.3	25
Mini drum with recycled irrigation water	Artificially inoculated	*L. monocytogenes*	2.5	6	6	120
Glass jars	Artificially inoculated	*L. monocytogenes*	2.5	5.5	7	146

constantly irrigated with water, which gathers microorganisms as it passes through the sprouting seeds. This spent irrigation water provides a better representation of the microbial population in the production batch. The sampling and microbial testing of spent irrigation water are relatively simple compared to those of sprouts. In addition, the sizes of the water samples can be easily scaled up to increase the chance of detecting the pathogens. Populations of *Salmonella* and *E. coli* O157:H7 in spent irrigation water are highly correlated with the populations on the germinating seeds (naturally contaminated or artificially inoculated), with populations approximately 1 $\log_{10}$ lower in the spent irrigation water (64, 82, 160). When alfalfa seeds naturally contaminated with *Salmonella* were sprouted, the numbers of confirmed positive samples were the same for the sampled sprouts and the spent irrigation water (161), suggesting that water analysis may accurately indicate the presence of pathogens on the sprouts.

Detailed protocols for the sampling and microbial testing of spent irrigation water and sprouts for the pathogens *Salmonella* and *E. coli* O157:H7 were included in the 1999 FDA guidance documents for the commercial production of sprouts (49, 157). The FDA recommended that, whenever possible, spent irrigation water rather than sprouts be sampled and that each production batch be sampled independently. Briefly, 1-liter samples of spent irrigation water should be aseptically collected as early as 48 h into the sprouting process and then subjected to microbiological testing with inclusion of an enrichment step. For instances in which collection of spent irrigation water is not feasible (e.g., for sprouts grown in soil), the sprouts should be tested. They should be aseptically sampled at different locations in the drum or growing trays and homogenized in a blender or stomacher prior to analysis. Testing by an independent, certified laboratory is preferred.

The perishable nature of the sprouts requires a quick turnaround of test results, and the use of rapid pathogen test kits was recommended by the FDA (49). Although many rapid test kits were commercially available at the time the FDA guidance was issued, few had undergone the extensive collaborative testing required for official approval. For *Salmonella*, the Assurance Gold enzyme immunoassay (Biocontrol, Inc., Bellevue, Wash.; AOAC official method 999.08) and the Visual Immunoprecipitate VIP immunoassay (Biocontrol, Inc.; AOAC official method 999.09) were available. For *E. coli* O157, only the Visual Immunoprecipitate VIP enterohemorrhagic *E. coli* assay (Biocontrol, Inc.) was officially approved (AOAC official method 996.09), but additional experience specifically with sprout testing by researchers using the Reveal for *E. coli* O157:H7 (Neogen Corporation, Lansing, Mich.) provided support for including it as an alternative test in the FDA guidance document (49). Despite the high microbial levels in sprouts and spent irrigation water, pathogen enrichment is necessary even for the rapid test kit procedures, and there have been no direct (i.e., nonenrichment) methods validated. The FDA guidance (49) detailed the enrichment steps for *E. coli* O157:H7 (enrichment in modified buffered peptone water supplemented with acriflavin, cefsulodin, and vancomycin) and for *Salmonella* (preenrichment in buffered peptone water supplemented with novobiocin, followed by enrichments in Rappaport-Vassiliadis and tetrathionate broths).

The efficacy of these rapid test kits for detection of *E. coli* O157:H7 and *Salmonella* in sprout irrigation water has been evaluated, and some kits have been compared with official conventional culture methods as described in the FDA's *Bacteriological Analytical Manual* (BAM) (47). The Reveal for *E. coli* O157:H7 and the VIP enterohemorrhagic *E. coli* tests could detect the presence of *E. coli* O157:H7 in 100 and 72%, respectively, of 36 samples inoculated at levels of 0.6 to 3.6 CFU/ml (182). The Assurance Gold for *Salmonella* enzyme immunoassay and the VIP for *Salmonella* test successfully detected 100% of all 66 samples inoculated with 0.67 to 3.6 CFU of the pathogen/ml, whereas the BAM method detected *Salmonella* in only 49 of the 66 inoculated samples (174).

In addition to those recommended by the FDA, many other methods have been evaluated for pathogen testing of sprouts. Most of the methods evaluated have targeted *Salmonella* and *E. coli* O157 (the two most common pathogens involved in illness outbreaks from sprouts) and may be categorized as conventional culture methods, immunoassay-based methods, and nucleic acid-based methods. The culture methods include the FDA BAM protocol (174), the use of various enrichment and recovery media (119, 182, 189), and membrane filter plating on selective agar media (160). The immunoassay-based methods include the use of several commercial lateral flow devices (e.g., VIP, Reveal, and Quix), an antibody-direct epifluorescent filter (Ab-DEFT) technique (160), an immunomagnetic capture and time-resolved fluorescence method (175), an antibody-based fiber-optic evanescent-wave biosensor system (94), the use of a multiarray-based immunosensor (165), and an electrochemical sandwich immunoassay (114). Nucleic acid-based tests evaluated include the GENE-TRAK *Salmonella* direct labeled probe (DLP) assay (162), the TaqMan *E. coli* O157:H7 PCR assay (62), multiplex PCR (63, 98), the BAX PCR assay (152, 162, 163), and real-time PCR (101). Table 3 summarizes the performance of most published methods with respect to the enrichment conditions used, the recovery protocol used, and the limit of detection.

The fiber-optic-based biosensor system relies on a polyclonal antibody for capturing *Salmonella* in spent irrigation water and then a monoclonal antibody for detection (94). When assays were done at 67 h into the sprouting

Table 3 Methods for detection of pathogens in sprouts, spent irrigation water, and seeds

Method(s)	Pathogen	Sample tested	Enrichment conditions[a]	Recovery improvement step(s)[b]	Detection limit	Reference
Culture based						
BAM protocols	*Salmonella* spp.	Spent irrigation water (alfalfa)	Preenrichment in lactose broth, 35°C, 22–26 h; selective enrichment in RV&TT, 42°C, 22–26 h	Plating on XLD, BS, HE	Positive for 39 of 66 samples inoculated at levels of 0.6–3.6 CFU/ml	174
	Salmonella spp.	Spent irrigation water and sprouts (alfalfa)	Preenrichment in lactose broth, 35°C, 22–26 h; selective enrichment in RV&TT, 42°C, 22–26 h	Plating on XLD, BS, HE	Positive for 100% of 64 samples inoculated at levels of 1–180 CFU/g	162
	E. coli O157:H7	Radish sprouts	Enrichment in mEC + n, 42°C, 18 h	IMS followed by plating on TC-SMAC agar, BCM 0157 agar, and CHROMagar O157	Positive for more than 90% of 80 samples inoculated at a level of 20.4 CFU/25 g	119
	E. coli O157:H7	Spent irrigation water (alfalfa)	Enrichment in mBPW + ACV at 42°C and 140 rpm for 24 h and EEB (1/4 concn of cefixime) at 37°C and 140 rpm for 24 h	Plating on TC-SMAC	Positive for 50 to 58% of 36 samples inoculated at levels of 0.41–1.34 CFU/ml	182
	E. coli O157:H7	Spent irrigation water (alfalfa)	Enrichment in mBPW + ACV at 42°C and 140	IMS and plating on TC-SMAC	Positive for 100% of 36 samples inoculated at	182

			rpm for 24 h		levels of 0.41–1.34 CFU/ml	
	E. coli O157:H7	Alfalfa sprouts	Enrichment in EEB, EEB (1/4 concn of cefixime), mEC at 37°C, 24 h with agitation; mBPW or mBPW + ACV, 42°C, 24 h with agitation	IMS and plating on TC-SMAC	Positive for 13 to 14 of 15 samples inoculated at levels of 0.12–0.42 CFU/g	183
	E. coli O157:H7	Alfalfa seeds	Enrichment in mTSB	Soaking of seeds for 1 h; IMS postenrichment		189
Membrane filter plating on BCM	*E. coli* O157:H7	Spent irrigation water (alfalfa)	ND	Plating on selective medium (BCM O157)	Not determined	160
	E. coli O26	Radish sprouts	mEC + n, 42°C, 18 h	IMS; plating on Rainbow agar O157 plus novobiocin	Positive for 9 of 9 samples inoculated at a level of 5.9 CFU/25 g	77
	E. coli O26	Alfalfa sprouts	Enrichment in mEC + n, 42°C, 18 h	IMS and plating on cefixime-tellurite O26 medium	Positive for 5 of 10 samples inoculated at a level of 2 CFU/25 g	78

(continued)

Table 3 Methods for detection of pathogens in sprouts, spent irrigation water, and seeds *(continued)*

Method(s)	Pathogen	Sample tested	Enrichment conditions[a]	Recovery improvement step(s)[b]	Detection limit	Reference
Immunoassay based						
Lateral flow device VIP	*Salmonella* spp.	Spent irrigation water (alfalfa)	Preenrichment in BPW + n, 35°C, 18–26 h; selective enrichment in RV&TT, 42°C, 5–8 h; post-enrichment in TSB + n, 42°C, 16–20 h	ND	Positive for 100% of 66 samples inoculated at levels of 0.6–3.6 CFU/ml	174
	E. coli O157:H7	Spent irrigation water and sprouts (alfalfa)	ND	ND	6–7 $\log_{10}$ CFU/g within 20 min	160
	E. coli O157:H7	Spent irrigation water (alfalfa)	Enrichment in mBPW + ACV at 42°C, 140 rpm, 24 h; EEB (1/4 concn of cefixime) at 37°C, 140 rpm, 24 h	ND	Positive for 72 to 75% of 36 samples inoculated at levels of 0.41–1.34 CFU/ml	182
	E. coli O157:H7	Alfalfa sprouts	Enrichment in mBPW + ACV, 42°C, 24 h with agitation (most effective enrichment)	ND	Positive for 9 of 15 samples inoculated at levels of 0.12–0.42 CFU/g	183
Lateral flow device Reveal	*E. coli* O157:H7	Spent irrigation water and sprouts (alfalfa)	ND	ND	5–6 $\log_{10}$ CFU/g within 20 min	160

	E. coli O157:H7	Spent irrigation water (alfalfa)	Enrichment in mBPW + ACV, 42°C, 140 rpm, 24 h; EEB (1/4 concn of cefixime), 37°C, 140 rpm, 24 h	ND	Positive for 100% of 36 samples inoculated at levels of 0.41–1.34 CFU/ml	182
Lateral flow device Quix	*E. coli* O157:H7	Spent irrigation water (alfalfa)	No enrichment or enrichment in BHI at room temp or 37°C	ND	Cannot be determined	62
Enzyme-linked immunosorbent assay Assurance Gold	*Salmonella* spp.	Spent irrigation water (alfalfa)	Preenrichment in BPW + n, 35°C, 18–26 h; selective enrichment in RV&TT, 42°C, 5–8 h; postenrichment in TSB + n, 42°C, 16–20 h	ND	Positive for 100% of 66 samples inoculated at levels of 0.6–3.6 CFU/ml	174
	Salmonella spp.	Spent irrigation water and sprouts (alfalfa)	Preenrichment in BPW + n, 35°C, 18–26 h; selective enrichment in RV&TT, 42°C, 5–8 h; postenrichment in TSB + n, 42°C 16–20 h	ND	Positive for 100% of 64 samples inoculated at levels of 1–180 CFU/g	174
Time-resolved fluorescence-based immunoassay	*Salmonella* spp.	Spent irrigation water and sprouts (alfalfa)	Enrichment in BHI at 37°C, 4 h, 160 rpm	IMS	Cannot be determined	175

(continued)

Table 3 Methods for detection of pathogens in sprouts, spent irrigation water, and seeds *(continued)*

Method(s)	Pathogen	Sample tested	Enrichment conditions[a]	Recovery improvement step(s)[b]	Detection limit	Reference
	E. coli O157:H7	Spent irrigation water and sprouts (alfalfa)	Enrichment in BHI, 37°C, 4 h, 160 rpm	IMS	Cannot be determined	175
Ab-DEFT	*E. coli* O157:H7	Spent irrigation water and sprouts (alfalfa)	ND	Membrane filtration	3.5 $\log_{10}$ CFU/g within 30 min	160
Evanescent wave-based multianalyte array biosensor	*S. enterica* serovar Typhimurium	Sprout homogenate and spent irrigation water (alfalfa)	ND	ND	6.6 $\log_{10}$ CFU/g (sprout) or 5.6 $\log_{10}$ CFU/ml (rinse water) within 15 min	165
Fiber-optic biosensor (RAPTOR)	*S. enterica* serovar Typhimurium	Spent irrigation water (alfalfa)	ND	ND	Cannot be determined	94
Fiber-optic biosensor (Analyte 2000)	*E. coli* O157:H7	Spent irrigation water (alfalfa)	ND	ND	Consistent positives only for samples inoculated at a level of 6.0 $\log_{10}$ CFU/ml	102
Disposable electrochemical biosensor	*E. coli* O157:H7	Alfalfa sprout homogenate	ND	ND	81 CFU/ml (within 6 min)	114
Nucleic acid based						
PCR-based TaqMan *E. coli* O157:H7 assay	*E. coli* O157:H7	Spent irrigation water (alfalfa)	No enrichment and enrichment in BHI, 37°C, 18 h	ND	Cannot be determined	62

BAX PCR	*Salmonella* spp.	Spent irrigation water and sprouts (alfalfa)	Preenrichment in lactose broth, 35°C, 22–26 h; postenrichment in BHI, 35°C, 3 h	ND	Positive for 90.6% of 64 samples inoculated at levels of 1–180 CFU/g	162
	S. enterica serovar Enteritidis	Alfalfa sprouts	Preenrichment in BPW, 37°C, overnight; postenrichment in BHI, 3 h, 37°C	ND	Positive for 5 of 6 samples inoculated at 1 CFU/25 g	152
	S. enterica serovar Enteritidis	Alfalfa sprouts	Preenrichment in mTSB, 37°C, 24 h, 150 rpm; postenrichment in BHI, 37°C, 3 h	ND	Positive for 12 of 12 samples inoculated at a level of 10 CFU/25 g	163
	S. enterica serovar Typhimurium	Alfalfa sprouts	Preenrichment in lactose, 37°C, 24 h; postenrichment in BHI, 37°C, 3 h	ND	Positive for 4 of 9 samples inoculated at a level of 4 CFU/25 g	102
	Salmonella spp.	Alfalfa seeds	Enrichment in BPW, 8 h, 37°C	IMS	Positive for 4 or 5 samples inoculated with 2 CFU of untreated *Salmonella*/25g; positive for 3 or 5 samples inoculated with 2–3 CFU of heat-injured *Salmonella*/25 g	100

(continued)

Table 3 Methods for detection of pathogens in sprouts, spent irrigation water, and seeds *(continued)*

Method(s)	Pathogen	Sample tested	Enrichment conditions[a]	Recovery improvement step(s)[b]	Detection limit	Reference
	E. coli O157:H7	Alfalfa sprouts	Enrichment in EEB, 24 h with shaking, 37°C	ND	Positive for 1 of 6 samples inoculated at a level of 10 CFU/25 g	152
	E. coli O157:H7	Alfalfa sprouts	Preenrichment in mTSB, 37°C, 24 h, 150 rpm; postenrichment in mEC, 42°C, overnight at 180 rpm	ND	Positive for 23 of 23 samples inoculated at a level of 10 CFU/25 g	163
	L. monocytogenes	Alfalfa sprouts	Preenrichment in Demi-Fraser broth, 20°C, 22–24 h; selective enrichment in MOPS-BLEB, 37°C, 20–24 h	ND	Positive for 4 of 6 samples inoculated at 1 CFU/25 g	152
	L. monocytogenes	Alfalfa sprouts	Preenrichment in Demi-Fraser broth, 30°C, 22–24 h; selective enrichment in MOPS-BLEB, 37°C, 20–24 h	ND	Positive for 12 of 12 samples inoculated at a level of 10 CFU/25 g	163
Molecular beacon real-time PCR	*S. enterica* serovar Typhimurium	Alfalfa sprouts	Enrichment in BPW, 37°C, 18 ± 2 h	ND	Positive for 5 of 9 samples inoculated at a level of 4 CFU/25 g	102

Multiplex PCR	*E. coli* O157:H7		Enrichment in mEC + n, 37°C, 24 h, 150 rpm	IMS; use of PrepMan for DNA extraction	1 CFU/g	63
	E. coli O157:H7 and *Salmonella* and *Shigella* spp.	Alfalfa sprouts	Enrichment in BHI, 37°C, 24 h	ND	1.9 $\log_{10}$ CFU/g	99

[a]RV&TT, Rappaport-Vassiliadis and tetrathionate broths; mEC + n, modified EC broth plus novobiocin; mBPW, modified buffered peptone water; ACV, acriflavin-cefsulodin-vancomycin; EEB, enterohemorrhagic *E. coli* enrichment broth; mEC, modified EC broth; mTSB, modified trypticase soy broth; ND, not done; BPW + n, buffered peptone water plus novobiocin; TSB + n, trypticase soy broth plus novobiocin; BHI, brain heart infusion; BPW, buffered peptone water; MOPS-BLEB, 3-[*N*-morpholino]propanesulfonic acid-buffered *Listeria* enrichment broth.

[b]XLD, xylose-lysine-desoxycholate agar; BS, bismuth sulfite agar; HE, Hektoen enteric agar; IMS, immunomagnetic separation.

process, *Salmonella* could be detected in 20 min without an enrichment step in the spent irrigation water after germination of alfalfa seeds inoculated with as few as 50 CFU of the pathogen/g (94). Unfortunately, the researchers failed to indicate the level of pathogens in the tested water, and thus it is difficult to estimate the detection limit of the biosensor. A similar optic-based biosensor, the Analyte 2000, was evaluated for detection of *E. coli* O157:H7 in sprout irrigation water (102). Although the sensor was able to detect *E. coli* O157:H7 in inoculated sprout irrigation water at levels as low as 1 to 10 CFU/ml, signal strength varied greatly among probes tested with the same samples. Consistent, positive reactions occurred only for samples containing *E. coli* O157:H7 at levels above 6 $\log_{10}$ CFU/ml. False positives were frequently observed. Taitt et al. (165) evaluated an evanescent wave-based, multianalyte array biosensor for detection of *Salmonella* in a variety of food matrices. This biosensor employed a 15-min sandwich immunoassay protocol and gave a detection limit of 5.6 $\log_{10}$ CFU/ml in spiked sprout rinse.

Difficulties in determining the sensitivities of published methods also exist in other cases. For example, it was reported that the time-resolved fluorescence-based immunoassay (175) could detect the presence of *E. coli* O157:H7 in seeds inoculated at a level of 4 CFU/g. However, the actual detection procedure was performed on the spent irrigation water collected after allowing the inoculated seeds to sprout for 48 h. Because the sprouting process may encourage pathogen proliferation, the level of *E. coli* O157:H7 after 48 h of sprouting may have exceeded 4 CFU/g; therefore, it gives a false estimate of assay sensitivity to state the seed inoculation level but not the level of pathogen in the spent irrigation water tested. Failure to state the level of pathogen present in the spent irrigation water samples collected for testing also makes it impossible to determine the sensitivity of the TaqMan PCR assay and the Quix *E. coli* O157 sprout assay (62).

In only a few reports were these experimental methods directly compared with the FDA-recommended rapid immunoassays or with official BAM methods (47). Stewart et al. (160) compared the capabilities to detect *E. coli* O157:H7 of the two FDA-recommended rapid immunoassays and the following methods: direct plating of enrichment preparations onto tellurite-cefixime-sorbitol MacConkey (TC-SMAC) and BCM O157:H7(+) agars; membrane filtration of spent irrigation water and incubation of the filters on BCM O157:H7(+) agar; and Ab-DEFT analysis of sprouts and spent irrigation water. Results were similar for all methods. In one study, the nucleic acid-based BAX PCR assay for *Salmonella* was equivalent in sensitivity to culture methods for detection of *Salmonella* inoculated onto alfalfa sprouts and reduced the overall testing period by a minimum of 2 days (152).

However, this study indicated that both the BAX PCR assay for *E. coli* O157:H7 and the FDA BAM culture methods were unable to consistently detect *E. coli* O157:H7 from inoculated alfalfa sprouts, presumably due to competition by native microflora during preenrichment or release of toxic plant components. The very high level of native microflora on sprouts was also suggested to be responsible for the relatively low detection sensitivity for *Salmonella, E. coli* O157:H7, and *Shigella* spp. compared to the detection sensitivity with other types of produce subjected to a multiplex PCR assay after nonselective enrichment (98). Another study indicated that the GENE-TRAK *Salmonella* DLP, the Assurance Gold for *Salmonella,* and the FDA BAM methods gave comparable results for detecting *Salmonella* from either naturally contaminated or inoculated sprouts and spent irrigation water but the BAX PCR assay for *Salmonella* was not as sensitive (162).

Further improvement of recovery of pathogens from sprouts may be achieved by the use of optimal processing methods and nonselective enrichment broths, by the use of immunomagnetic capture, and by modification of selective agar media. When three methods (washing, stomaching, and homogenizing in a rotor-stator homogenizer) were compared for processing inoculated alfalfa sprouts, stomaching resulted in slightly higher recoveries than the other two methods (17). Enrichment in buffered peptone water containing 0.5% sodium thioglycolate led to better recovery of unstarved and starved cells of EC O157 from inoculated radish sprouts than did enrichment in modified EC broth plus novobiocin or buffered peptone water alone (144). The use of modified EC broth plus novobiocin was also found to be detrimental for recovery of injured *E. coli* O157:H7 inoculated onto radish sprouts (76). Concentration of *E. coli* O157:H7 and *E. coli* O26 from enrichment broth by immunomagnetic separation techniques enhanced the sensitivity and specificity of detection (75, 77, 119, 183, 189). Improved differentiation of *E. coli* O157 from bacteria naturally present on radish sprouts was obtained by the addition of salicin and 4-methylumbelliferyl–β-galactopyranoside to the selective agar medium TC-SMAC (66).

Methods have also been devised for the isolation of parasites from sprouts. Mung bean sprouts were reported to have a higher occurrence of contamination with *Cryptosporidium* and *Giardia* than other types of produce in Norway (140). Oocysts of *Cryptosporidium* and *Cyclospora* spp. as well as cysts of *Giardia* were detectable on artificially inoculated mung bean sprouts by a combination of washing, sonication, and immunomagnetic separation (139, 141).

Testing each production batch for pathogens according to FDA recommendations is arguably the single most effective way to enhance the microbiological safety of sprouts. However, microbial testing during sprout production has not been fully implemented by the industry. In a survey of California sprout growers to determine industry compliance with the FDA guidance (170), 67% of the growers surveyed indicated that testing of spent irrigation water according to the FDA guidance was performed, even though 94% of the firms surveyed had read and reviewed the guidance. The noncompliance may be due in part to the extra cost required for the testing. Some producers have proposed pooling samples from multiple batches to lower the testing cost. The FDA is concerned that pooling will reduce the sensitivity of tests due to the dilution of the level of pathogen in a contaminated sample with samples that are not contaminated (157). One way to alleviate this concern is to institute a sample preconcentration step that would allow the entire pooled sample to be analyzed by a single test. For example, a 10-liter pooled sample of spent irrigation water from 10 production batches (1 liter from each batch) would be concentrated to an appropriate volume for analysis. A tangential flow filtration system has been developed that would allow the concentration of 10 liters of spent irrigation water into 100 ml within 2 h (64). Since sample concentration also results in an increase in the level of background microflora and other potentially interfering substances, the efficacy of the recommended rapid tests for detection of *E. coli* O157:H7 and *Salmonella* in the concentrated spent irrigation water was evaluated. It was found that the Assurance Gold enzyme immunoassay and the Reveal assay were able to detect the presence of 1 CFU of *Salmonella* and *E. coli* O157:H7 in 10 ml of concentrated sample (which corresponds to 1 liter of sprout water sample prior to filtration) (T. Fu, unpublished results). These results suggested that, with the addition of a sample preconcentration step, the entire pooled sample may be analyzed with a single test while the sensitivity of the test is maintained.

Pathogen Sampling and Detection in Seeds

There is a need to develop methods for seed testing, especially when large numbers of samples need to be tested quickly, for example, in conducting trace backs in outbreak investigations. The isolation of bacterial pathogens from seeds has been a subject of several research efforts.

Contamination of seeds by pathogens appears to be nonhomogeneous and to take place at low levels, and recovery is difficult. Despite the fact that *Salmonella* and *E. coli* O157:H7 have been shown to survive on artificially and naturally contaminated seeds for up to 2 years or more depending on the storage conditions (11, 68, 90, 95, 100, 166, 189), pathogens often were not isolated from the seed lots implicated in illness outbreaks (14, 15, 117). Nevertheless, when successful, seed testing has played an important role in outbreak investigations (84, 85).

Although there is some evidence that individual seeds in naturally contaminated lots may harbor pathogen populations of greater than 4 $\log_{10}$ CFU (169), other studies have estimated populations of *Salmonella* in naturally contaminated seed lots to range from 0.07 most probable number (MPN)/100 g to approximately 10 MPN/100 g (1, 84, 85, 100). With the often low overall contamination level, a sample size larger than the standard 25 g is needed for testing of outbreak-associated seeds (164). Inadequate sampling and small sample sizes have been suggested as the reasons for the failure of laboratory analyses in detecting *E. coli* O157:H7 and *Salmonella* in implicated seed lots (16). Testing of sprouting seed may increase the chance of finding target pathogens. Evidence exists that salmonellae can be cultured from naturally contaminated seeds only after sprouting (130).

Inami et al. (85) compared two processing methods for recovery of *Salmonella* from naturally contaminated alfalfa seeds. Seeds were either sprouted or shredded in a blender before preenrichment and culture. Results indicated that the two methods were comparable for detection of the pathogen. Liao and Fett (100) reported that two sequential preenrichment steps in buffered peptone water, rather than a single preenrichment step, prior to selective enrichment led to a higher detection rate for *Salmonella* from naturally contaminated seeds, indicating the possible presence of injured bacterial cells. Suslow et al. (164) found that the sensitivity of detection of *S. enterica* serovar Mbandaka from seeds associated with a multistate salmonellosis outbreak was improved by a combination of nonselective and selective enrichment, followed by immunocapture and plating or immunocapture in combination with PCR. Liao and Shollenberger (99) found that of five indicator agar media compared for the detection of *Salmonella* in the presence of native seed microflora, modified semisolid Rappaport-Vassiliadis medium was the most sensitive. Wu et al. (189) reported that the optimal procedure for isolation of *E. coli* O157:H7 from artificially inoculated alfalfa seeds was to first soak seeds for 1 h in sterile water, pummel for 1 min, enrich in modified tryptic soy broth, concentrate the cells by immunocapture, and plate onto selective agar media (Table 3). The use of immunomagnetic beads was also recommended by Liao and Shollenberger (99) to eliminate substances in seed homogenates that are inhibitory of PCR (Table 3). By combining immunomagnetic separation with the BAX PCR system for screening *Salmonella*, as few as 2 to 5 CFU of heat-injured *Salmonella* in 25 g could be detected within 24 h. The parasites *Cryptosporidium* and *Giardia* have also been isolated from naturally contaminated seeds, and on the basis of research results, testing of seeds rather than spent irrigation water was recommended for detection of these two parasites (142).

Seed screening has been proposed as an additional step in a multiple-hurdle approach to prevent sprout-associated outbreaks (143). Proper sampling, inspection, and testing of seeds for pathogens can substantially reduce the chance of using contaminated seeds for sprout production and therefore help prevent foodborne illness. However, improper seed screening protocols may still result in outbreaks (71).

A comprehensive seed-sampling and testing procedure has been developed by the sprouting industry (87). The procedure consists of a six-step process: sampling of seeds in bags that pass initial inspection (e.g., no evidence of rodent urine or bird droppings, and no holes in the bags), seed inspection under magnification, sprouting of seeds, spent water sampling, enrichment of sampled water, and pathogen testing. Briefly, the procedure calls for a sampling of at least 3 kg of seeds from each seed lot with 25-g subsamples from each bag. Assuming the level of pathogens found in seeds to be as low as 4 CFU per kg (117) and assuming uniform distribution of pathogens in seeds, a seed sample of 3 kg would give a probability of 99.9994% of finding a single pathogen cell in a 20-ton lot of seed (87). The entire sample is sprouted without prior sanitization by using commercial sprout production methods in an area segregated from commercial sprout production and at temperatures that support maximal pathogen growth if any pathogens are present. At 48 h of sprouting, a sample of the runoff water is collected and tested for pathogens by using FDA-recommended procedures (49). The adaptation of these seed-screening procedures has already prevented at least four potential sprout-related outbreaks due to *Salmonella* and *E. coli* O157:H7 (86).

REDUCING THE RISK OF FOODBORNE ILLNESS

Because of the exponential growth of microorganisms that occurs as a normal part of sprouting, measures must be taken to prohibit contamination both before and during production. It is generally believed that most of the outbreaks linked to sprouts have been due to the use of contaminated seeds. However, pathogens may enter the process via water, equipment, or handling and may present a hazard even if uncontaminated seeds are used. Thus, it is recognized that the risk of pathogen entry will be reduced by having in place not only good agricultural practices (GAPs) for seed production but also good manufacturing practices (GMPs) in the sprouting facility. Although sprout growers may be able to implement GMPs within their own facilities, adherence to the primary aspects of GAPs in producing seeds for sprouting is generally not within their control. Therefore, the potential presence of pathogens must be assumed, and sprout growers should also follow a program of seed disinfection and pathogen testing during production to enhance the safety of their product. The food safety challenge presented by the unique nature of sprouts requires a system of complementary risk reduction measures, i.e., an integrated pathogen management approach.

GAPs

Seeds that are commonly used for sprouting are generally not intended for human consumption. Alfalfa seeds, for example, are produced primarily to sow a pasture or cover crop. They have typical characteristics of a raw agricultural product, including exposure to fecal contamination from wild and domestic animals, manure, and soil. Their use for human food purposes is merely incidental to their primary function in the agricultural sector. Nevertheless, seeds for sprouting are transformed into human food in a very brief process, and so it is FDA policy to regulate the seeds themselves as food; i.e., the seeds are not exempt from sanitation inspections and seizure and condemnation if they are adulterated (46).

The FDA has issued recommendations for the production and distribution of seeds for sprouting (49) which call for following the GAPs that are pertinent to the growing of fresh fruits and vegetables (48). The purpose of these comprehensive measures is to minimize the likelihood of contamination by pathogens, and they are to be tailored for specific growing, harvesting, and packing procedures as appropriate for the various crops. Among the recommendations that apply to seeds for sprouting are reduction of microbial hazards from the use of manures and biosolids from municipal waste treatment, consideration of water sanitary quality, attention to worker hygiene, and pest control in packing and distribution. Contaminated seed conditioning (cleaning) equipment can lead to cross-contamination of seed lots. Such equipment should be thoroughly cleaned and sanitized between seed lots destined for sprouting for human consumption.

Even if the seeds have been produced under GAPs, mixing with other lots during seed distribution or sprouting may occur. The ability to conduct trace backs in the event of an illness outbreak may limit the number of illnesses by allowing regulators to identify seed lots that have been implicated in the outbreak. Therefore, trace back is recognized as an important risk reduction strategy and complement to GAPs. Trace back was elevated from recommended to mandatory with the passage of the Public Health Security and Bioterrorism Preparedness and Response Act of 2002 (57), which requires food facilities to establish and maintain records of the immediate sources and immediate recipients of food (i.e., "one up, one down"). Seed distributors and sprout growers are required to maintain these records.

GMPs, Seed Treatment, and Pathogen Testing

The food industry has long been required to follow GMPs in manufacturing, packaging, or holding food, and the FDA reminded the sprout industry of this commitment in its guidance documents (49). Current GMPs include sanitary operation, worker hygiene, maintenance of facilities and equipment so as to protect against contamination, and control in production, processing, warehousing, and distribution (56).

In addition to GMPs, the FDA guidance documents made two further recommendations: (i) the use of an approved treatment for reducing pathogens on the seeds immediately before sprouting and (ii) testing of each production batch for *Salmonella* and *E. coli* O157:H7. The preferred test material is the spent irrigation water, i.e., the water that has flowed over and through the sprouts during production. If collection of the water is impractical due to methodology or equipment constraints, then the sprouts themselves may be tested. Such end-product testing for pathogens is generally considered to be an ineffective safety assurance technique for most foods because of the low levels and sporadic nature of contamination. Seed sprouting, however, uniquely provides enrichment conditions leading to high population levels of the microbial flora; thus, sampling problems associated with sporadic, low-level contamination are less likely to occur. In fact, for sprouting, pathogen testing is an extremely effective technique and one proven by the sprout industry as a means to avoid distribution of contaminated product (86). A recent Monte Carlo simulation model using the available literature predicted that sprout sampling and spent irrigation water sampling would be more effective in the detection of pathogens than seed sampling prior to production (113).

HACCP Systems and Integrated Pathogen Management

Traditionally, hazard analysis and critical control point (HACCP) systems provide one or more critical control points in a food manufacturing process that ensure safety of the product. Although this key element is missing for sprout production (no treatment has been shown to effectively control pathogen growth), sprout growers have adopted elements of HACCP systems to enhance safety (88). Retail establishments that grow sprouts for direct sale to consumers must employ a HACCP plan according to the FDA Food Code (51, 58).

Sprout growers have advocated an integrated risk reduction approach involving not only seed sanitization and pathogen testing of spent irrigation water from production batches but also a seed certification and screening program. Acceptability of seeds prior to production sprouting would be determined by inspecting the seed shipment for evidence of filth and sampling the seeds according to a statistically based sampling plan. A composite seed sample is sprouted, and pathogen testing of the spent irrigation water by FDA-recommended procedures is performed. This seed-screening procedure does not preclude the necessity of testing each production batch according to the FDA guidance. All of these measures, applied together, are the best available options for reducing the microbiological risks associated with sprouts.

ACKNOWLEDGMENTS

We thank David Douds, Sam Palumbo, and Ethan Solomon for their critical reviews of the manuscript.

The mention of trade names or commercial products in this chapter is solely for the purpose of providing specific information and does not imply recommendation or endorsement by the U.S. Department of Agriculture or by the FDA.

REFERENCES

1. **Abbo, S., and D. L. Baggesen.** 1997. Growth of *Salmonella* Newport in naturally contaminated alfalfa sprouts and estimation of infectious dose in a Danish *Salmonella* Newport outbreak due to alfalfa sprouts. *Salmonella Salmonellosis* **97:**425–426.
2. **Andrews, W. H., C. R. Wilson, P. L. Poelma, A. Romero, and P. B. Mislivec.** 1979. Bacteriological survey of sixty health foods. *Appl. Environ. Microbiol.* **37:**559–566.
3. **Andrews, W. H., P. B. Mislivec, C. R. Wilson, V. R. Bruce, P. L. Poelma, R. Gibson, M. W. Trucksess, and K. Young.** 1982. Microbial hazards associated with bean sprouting. *J. Assoc. Off. Anal. Chem.* **65:**241–248.
4. **Ariefdjohan, M. W., P. E. Nelson, R. K. Singh, A. K. Bhunia, V. M. Balasubramaniam, and N. Singh.** 2004. Efficacy of high hydrostatic pressure treatment in reducing *Escherichia coli* O157 and *Listeria monocytogenes* in alfalfa seeds. *J. Food Sci.* **69:**M117–M120.
5. **Barak, J. D., L. C. Whitehand, and A. O. Charkowski.** 2002. Differences in attachment of *Salmonella enterica* serovars and *Escherichia coli* O157:H7 to alfalfa sprouts. *Appl. Environ. Microbiol.* **68:**4758–4763.
6. **Bari, M. L., E. Nazuka, Y. Sabina, S. Todoriki, and K. Isshiki.** 2003. Chemical and irradiation treatments for killing *Escherichia coli* O157:H7 on alfalfa, radish, and mung bean seeds. *J. Food Prot.* **66:**767–774.
7. **Bari, M. L., M. I. Al-Haq, T. Kawasaki, M. Nakauma, S. Todoriki, S. Kawamoto, and K. Isshiki.** 2004. Irradiation to kill *Escherichia coli* O157:H7 and *Salmonella* on ready-to-eat radish and mung bean sprouts. *J. Food Prot.* **67:**2263–2268.
8. **Becker, B., and W. H. Holzapfel.** 1997. Microbiological risk of prepacked sprouts and measures to reduce total counts. *Arch. Lebensmittelhygiene* **48:**81–84.
9. **Bennik, M. H. J., W. Vorstman, E. J. Smid, and L. G. M. Gorris.** 1998. The influence of oxygen and carbon dioxide on the growth of prevalent Enterobacteriaceae and *Pseudomonas* species isolated from fresh and controlled-atmosphere-stored vegetables. *Food Microbiol.* **15:**459–469.
10. **Beuchat, L. R.** 1997. Comparison of chemical treatments to kill *Salmonella* on alfalfa seeds destined for sprout production. *Int. J. Food Microbiol.* **34:**329–333.

11. **Beuchat, L.R., and A. J. Scouten.** 2002. Combined effects of water activity, temperature and chemical treatments on the survival of *Salmonella* and *Escherichia coli* O157:H7 on alfalfa seeds. *J. Appl. Microbiol.* **92:**382–395.
12. **Beuchat, L. R., T. E. Ward, and C. A. Pettigrew.** 2001. Comparison of chlorine and a prototype produce wash product for effectiveness in killing *Salmonella* and *Escherichia coli* O157:H7 on alfalfa seeds. *J. Food Prot.* **64:**152–158.
13. **Breidt, F., and H. P. Fleming.** 1997. Using lactic acid bacteria to improve the safety of minimally processed fruits and vegetables. *Food Technol.* **51:**44–49.
14. **Breuer, T., D. H. Benkel, R. L. Shapiro, W. N. Hall, M. M. Winnett, M. J. Linn, J. Neimann, T. J. Barrett, S. Dietrich, F. P. Downes, D. M. Toney, J. L. Pearson, H. Rolka, L. Slutsker, P. M. Griffin, and the Investigation Team.** 2001. A multistate outbreak of *Escherichia coli* O157:H7 infections linked to alfalfa sprouts grown from contaminated seeds. *Emerg. Infect. Dis.* **7:**977–982.
15. **Brooks, J. T., S. Y. Rowe, P. Shillam, D. M. Heltzel, S. B. Hunter, L. Slutsker, R. M. Hoekstra, and S. P. Luby.** 2001. *Salmonella* Typhimurium infections transmitted by chlorine-pretreated clover sprout seeds. *Am. J. Epidemiol.* **154:**1020–1028.
16. **Buck, J. W., R. R. Walcott, and L. R. Beuchat.** 21 January 2003, posting date. Recent trends in microbiological safety of fruits and vegetables. *Plant Health Prog.* [Online.] doi:10.1094/PhP-2003-0121-01-RV.
17. **Burnett, A. B., and L. R. Beuchat.** 2001. Comparison of sample preparation methods for recovering *Salmonella* from raw fruits, vegetables, and herbs. *J. Food Prot.* **64:**1459–1465.
18. **Cai, Y., L.-K. Ng, and J. M. Farber.** 1997. Isolation and characterization of nisin-producing *Lactococcus lactis* subsp. *lactis* from bean-sprouts. *J. Appl. Microbiol.* **83:**499–507.
19. **Callister, S. M., and W. A. Agger.** 1987. Enumeration and characterization of *Aeromonas hydrophila* and *Aeromonas caviae* isolated from grocery store produce. *Appl. Environ. Microbiol.* **53:**249–253.
20. **Campbell, R. C.** 2003. *Biological Control of Microbial Plant Pathogens.* Cambridge University Press, Cambridge, United Kingdom.
21. **Cao, G., E. Sofic, and R. L. Prior.** 1996. Antioxidant capacity of tea and common vegetables. *J. Agric. Food Chem.* **44:**3426–3431.
22. **Castro-Rosas, J., and E. F. Escartin.** 1999. Incidence and germicide sensitivity of *Salmonella typhi* and *Vibrio cholerae* O1 in alfalfa sprouts. *J. Food Safety* **19:**137–146.
23. **Castro-Rosas, J., and E. F. Escartin.** 2000. Survival and growth of *Vibrio cholerae* O1, *Salmonella typhi*, and *Escherichia coli* O157:H7 in alfalfa sprouts. *J. Food Sci.* **65:**162–165.
24. **Charkowski, A. O., C. Z. Sarreal, and R. E. Mandrell.** 2001. Wrinkled alfalfa seeds harbor more aerobic bacteria and are more difficult to sanitize than smooth seeds. *J. Food Prot.* **64:**1292–1298.
25. **Charkowski, A. O., J. D. Barak, C. Z. Sarreal, and R. E. Mandrell.** 2002. Differences in growth of *Salmonella enterica* and *Escherichia coli* O157:H7 on alfalfa sprouts. *Appl. Environ. Microbiol.* **68:**3114–3120.
26. **Cooley, M. B., W. G. Miller, and R. E. Mandrell.** 2003. Colonization of *Arabidopsis thaliana* with *Salmonella enterica* and enterohemorrhagic *Escherichia coli* O157:H7 and competition by *Enterobacter asburiae. Appl. Environ. Microbiol.* **69:**4915–4926.
27. **Danielsson-Tham, M.-L., and K. Holm.** 1983. Bacteriological quality of sprouts and salads containing sprouts. *Var Föda* **35:**11–19.

28. **Delaquis, P. J., P. L. Sholberg, and K. Stanich.** 1999. Disinfection of mung bean seed with gaseous acetic acid. *J. Food Prot.* **62**:953–957.
29. **Dong, Y., A. L. Iniguez, B. M. M. Ahmer, and E. W. Triplett.** 2003. Kinetics and strain specificity of rhizosphere and endophytic colonization by enteric bacteria on seedlings of *Medicago sativa* and *Medicago truncatula*. *Appl. Environ. Microbiol.* **69**:1783–1790.
30. **Donlan, R. M., and J. W. Costerton.** 2002. Biofilms: survival mechanisms of clinically relevant microorganisms. *Clin. Microbiol. Rev.* **15**:167–193.
31. **Dychdala, G. R.** 1991. Chlorine and chlorine compounds, p. 131–151. *In* S. S. Block (ed.), *Disinfection, Sterilization, and Preservation,* 4th ed. Lea & Febinger, Philadelphia, Pa.
32. **Enomoto, K.** 2004. Use of bean sprout *Enterobacteriaceae* isolates as biological control agents of *Pseudomonas fluorescens*. *J. Food Sci.* **69**:FMS17–FMS22.
33. **Enomoto, K., T. Takizawa, N. Ishikawa, and T. Suzuki.** 2002. Hot-water treatments for disinfecting alfalfa seeds inoculated with *Escherichia coli* ATCC 25922. *Food Sci. Technol. Res.* **8**:247–251.
34. **Environmental Protection Agency.** 1998. 40 CFR parts 9, 141, and 142, national primary drinking water regulations: disinfectants and disinfection byproducts, final rule. *Fed. Regist.* **63**:69390–69476.
35. **Fahey, J. W., Y. Zhang, and P. Talalay.** 1997. Broccoli sprouts: an exceptionally rich source of inducers of enzymes that protect against chemical carcinogens. *Proc. Natl. Acad. Sci. USA* **94**:10367–10372.
36. **Fan, X., and D. W. Thayer.** 2001. Quality of irradiated alfalfa sprouts. *J. Food Prot.* **64**:1574–1578.
37. **Fan, X., D. W. Thayer, and K. J. B. Sokorai.** 2004. Changes in growth and antioxidant status of alfalfa sprouts during sprouting as affected by gamma irradiation of seeds. *J. Food Prot.* **67**:561–566.
38. **Fett, W. F.** 2000. Naturally occurring biofilms on alfalfa and other types of sprouts. *J. Food Prot.* **63**:625–632.
39. **Fett, W. F.** 2002. Reduction of the native microflora on alfalfa sprouts during propagation by addition of antimicrobial compounds to the irrigation water. *Int. J. Food Microbiol.* **72**:13–18.
40. **Fett, W. F.** 2002. Factors affecting the efficacy of chlorine against *Escherichia coli* O157:H7 and *Salmonella* on alfalfa seed. *Food Microbiol.* **19**:135–149.
41. **Fett, W. F.** 2002. Reduction of *Escherichia coli* O157:H7 and *Salmonella* spp. on laboratory-inoculated mung bean seed by chlorine treatment. *J. Food Prot.* **65**:848–852.
42. **Fett, W. F.** 2005. Interventions to ensure the microbiological safety of sprouts, p. 187–209. *In* G. M. Sapers, J. R. Gorny, and A. E. Yousef (ed.), *Microbiology of Fruits and Vegetables,* in press. CRC Press, Boca Raton, Fla.
43. **Fett, W. F., and P. H. Cooke.** 2003. Scanning electron microscopy of native biofilms on mung bean sprouts. *Can. J. Microbiol.* **49**:45–50.
44. **Fett, W. F., and P. H. Cooke.** 2003. Reduction of *Escherichia coli* O157:H7 and *Salmonella* on laboratory-inoculated alfalfa seed with commercial citrus-related products. *J. Food Prot.* **66**:1158–1165.
45. **Fett, W. F., and P. H. Cooke.** 2005. A survey of native microbial aggregates on alfalfa, clover, and mung bean sprout cotyledons for thickness as determined by confocal scanning laser microscopy. *Food Microbiol.* **22**:253–259.

46. **Food and Drug Administration.** 1989. *Seeds for Sprouting Prior to Food Use, i.e., Dried Mung Beans, Alfalfa Seeds, etc.* Compliance policy guide 7120.28, section 555.750. [Online.] Food and Drug Administration, Washington, D.C. http://www.fda.gov/ora/compliance_ref/cpg/cpgfod/cpg555-750.html.
47. **Food and Drug Administration.** 1998. *Bacteriological Analytical Manual,* 8th ed., revision A. AOAC International, Gaithersburg, Md.
48. **Food and Drug Administration.** 26 October 1998, posting date. *Guidance for Industry: Guide To Minimize Food Safety Hazards for Fresh Fruits and Vegetables.* [Online.] Food and Drug Administration, Washington, D.C. http://www.cfsan.fda.gov/~dms/prodguid.html.
49. **Food and Drug Administration.** 1999. Guidance for industry: reducing microbial food safety hazards for sprouted seeds and guidance for industry: sampling and microbial testing of spent irrigation water during sprout production. *Fed. Regist.* **64**:57893–57902.
50. **Food and Drug Administration.** 2000. Irradiation in the production, processing and handling of food. *Fed. Regist.* **65**:64605–64607.
51. **Food and Drug Administration.** 7 April 2004, revision date. [Online.] Food and Drug Administration, Washington, D.C. http://www.cfsan.fda.gov/~dms/fc01-1.html.
52. **Food and Drug Administration.** 2 October 2002, posting date. *Consumers Advised of Risks Associated with Eating Raw and Lightly Cooked Sprouts.* [Online.] Food and Drug Administration, Washington, D.C. http://www.cfsan.fda.gov/~lrd/tpsprout.html.
53. **Food and Drug Administration.** 19 February 2003, posting date. *Recalls and Field Corrections: Foods—Class I.* [Online.] Food and Drug Administration, Washington, D.C. http://www.fda.gov/bbs/topics/enforce/2003/ENF00783.html.
54. **Food and Drug Administration.** 19 August 2004, posting date. *Note to Firms That Grow, Condition, Store, or Distribute Seed for Sprouting and to Firms That Produce, Pack, or Ship Fresh Sprouts.* [Online.] Food and Drug Administration, Washington, D.C. http://www.cfsan.fda.gov/~dms/sproltr.html.
55. **Food and Drug Administration.** October 2004, posting date. *Produce Safety from Production to Consumption: 2004 Action Plan To Minimize Foodborne Illness Associated with Fresh Produce Consumption.* [Online.] Food and Drug Administration, Washington, D.C. http://www.cfsan.fda.gov/~dms/prodpla2.html.
56. **Food and Drug Administration.** 23 June 2004, posting date. *21 CFR Part 110—Current Good Manufacturing Practice in Manufacturing, Packing, or Holding Human Food.* [Online.] Food and Drug Administration, Washington, D.C. http://www.access.gpo.gov/nara/cfr/waisidx_04/21cfr110_04.html.
57. **Food and Drug Administration.** 2004. Establishment and maintenance of records under the Public Health Security and Bioterrorism Preparedness and Response Act of 2002. *Fed. Regist.* **69**:71561–71655.
58. **Food and Drug Administration.** December 2004, posting date. *Growing Sprouts in Retail Food Establishments.* [Online.] Food and Drug Administration, Washington, D.C. http://www.cfsan.gov/~dms/sprouret.html.
59. **Food and Drug Administration and Food and Drug Branch, California Department of Health Services.** 2003. *Safer Processing of Sprouts.* [Video.] Food and Drug Branch, California Department of Health Services, Sacramento, Calif.

60. **Food and Drug Administration and U.S. Department of Agriculture.** 23 December 2003, revision date. *Quantitative Assessment of Relative Risk to Public Health from Foodborne* Listeria monocytogenes *among Selected Categories of Ready-to-Eat Foods.* [Online.] Food and Drug Administration and U.S. Department of Agriculture, Washington, D.C. http://www.cfsan.fda.gov/~dms/lmr2-1.html.

61. **Forsberg, G.** 2004. Control of cereal seed-borne diseases by hot humid air seed treatment. Ph.D. thesis. Swedish University of Agricultural Sciences, Uppsala, Sweden.

62. **Fratamico, P. M., and L. K. Bagi.** 2001. Comparison of an immunochromatographic method and the Taqman *E. coli* O157:H7 assay for detection of *Escherichia coli* O157:H7 in alfalfa sprout spent irrigation water and in sprouts after blanching. *J. Ind. Microbiol. Biotechnol.* **27**:129–134.

63. **Fratamico, P. M., L. K. Bagi, and T. A. Pepe.** 2000. A multiplex polymerase chain reaction assay for rapid detection and identification of *Escherichia coli* O157:H7 in foods and bovine feces. *J. Food Prot.* **63**:1032–1037.

64. **Fu, T., D. Stewart, K. Reineke, J. Ulaszek, J. Schlesser, and M. Tortorello.** 2001. Use of spent irrigation water for microbiological analysis of alfalfa sprouts. *J. Food Prot.* **64**:802–806.

65. **Fu, T.-J., O. M. VanPelt, and K. F. Reineke.** 2003. Growth of *Salmonella* during sprouting of naturally contaminated alfalfa seeds as affected by sprouting conditions. *J. Food Prot.* **66**(Suppl. A):152, paper no. p268.

66. **Fujisawa, T., S. Sata, K. Aikawa, T. Takahashi, S. Yamai, and T. Shimada.** 2000. Modification of sorbitol MacConkey medium containing cefixime and tellurite for isolation of *Escherichia coli* O157:H7 from radish sprouts. *Appl. Environ. Microbiol.* **66**:3117–3118.

67. **Gagne, S., C. Richard, H. Rousseau, and H. Antoun.** 1987. Xylem-residing bacteria in alfalfa roots. *Can. J. Microbiol.* **33**:996–1000.

68. **Gandhi, M., S. Golding, S. Yaron, and K. R. Matthews.** 2001. Use of green fluorescent protein expressing *Salmonella* Stanley to investigate survival, spatial location, and control on alfalfa sprouts. *J. Food Prot.* **64**:1891–1898.

69. **Gandhi, M., and K. R. Matthews.** 2003. Efficacy of chlorine and calcinated calcium treatment of alfalfa seeds and sprouts to eliminate Salmonella. *Int. J. Food Microbiol.* **87**:301–306.

70. **Garland, J. L.** 1997. Analysis and interpretation of community-level physiological profiles in microbial ecology. *FEMS Microbiol. Ecol.* **24**:289–300.

71. **Gill, C. J., W. E. Keene, J. C. Mohle-Boetani, J. A. Farrar, P. L. Waller, C. G. Hahn, and P. R. Cieslak.** 2003. Alfalfa seed decontamination in a *Salmonella* outbreak. *Emerg. Infect. Dis.* **9**:474–479.

72. **Gill, J., and S. T. Abedon.** November. 2003, posting date. Bacteriophage ecology and plants. *APSnet* **2003**(Nov.). [Online.] http://apsnet.org/online/feature/phages/.

73. **Gorski, L., J. D. Palumbo, and K. D. Nguyen.** 2004. Strain-specific differences in the attachment of *Listeria monocytogenes* to alfalfa sprouts. *J. Food Prot.* **67**:2488–2495.

74. **Hallman, J., A. Quadt-Hallmann, W. F. Mahaffee, and J. W. Kloepper.** 1997. Bacterial endophytes in agricultural crops. *Can. J. Microbiol.* **43**:895–914.

75. **Hara-Kudo, Y., Y. Onoue, H. Konuma, H. Nakagawa, and S. Kumagai.** 1999. Comparison of enrichment procedures for isolation of *Escherichia coli* O157:H7 from ground beef and radish sprouts. *Int. J. Food Microbiol.* **50**:211–214.

76. **Hara-Kudo, Y., H. Konuma, H. Nakagawa, and S. Kumagai.** 2000. *Escherichia coli* O26 detection from foods using an enrichment procedure and an immunomagnetic separation method. *Lett. Appl. Microbiol.* **30:**151–154.
77. **Hara-Kudo, Y., M. Ikedo, H. Kodaka, H. Nakagawa, K. Goto, T. Masuda, H. Konuma, T. Kojima, and S. Kumagai.** 2000. Selective enrichment with a resuscitation step for isolation of freeze-injured *Escherichia coli* O157:H7 from foods. *Appl. Environ. Microbiol.* **66:**2866–2872.
78. **Hara-Kudo, Y., M. Ikedo, O. Komatsu, S. Yamamoto, and S. Kumagai.** 2002. Evaluation of a chromogenic agar medium for isolation of *Escherichia coli* O26. *Food Control* **13:**377–379.
79. **Harmon, S. M., D. A. Kautter, and H. M. Solomon.** 1987. *Bacillus cereus* contamination of seeds and vegetable sprouts grown in a home sprouting kit. *J. Food Prot.* **50:**62–65.
80. **Himathongkham, S., S. Nuanualsuwan, H. Riemann, and D. O. Cliver.** 2001. Reduction of *Escherichia coli* O157:H7 and *Salmonella* Typhimurium in artificially contaminated alfalfa seeds and mung beans by fumigation with ammonia. *J. Food Prot.* **64:**1817–1819.
81. **Holliday, S. L., A. J. Scouten, and L. R. Beuchat.** 2001. Efficacy of chemical treatments in eliminating *Salmonella* and *Escherichia coli* O157:H7 on scarified and polished alfalfa seeds. *J. Food Prot.* **64:**1489–1495.
82. **Howard, M. B., and S. W. Hutcheson.** 2003. Growth dynamics of *Salmonella enterica* strains on alfalfa sprouts and in waste seed irrigation water. *Appl. Environ. Microbiol.* **69:**548–553.
83. **Hu, H., J. J. Churey, and R. W. Worobo.** 2004. Heat treatments to enhance the safety of mung bean seeds. *J. Food Prot.* **67:**1257–1260.
84. **Inami, G., and S. E. Moler.** 1999. Detection and isolation of *Salmonella* from naturally contaminated alfalfa seeds following an outbreak investigation. *J. Food Prot.* **62:**662–664.
85. **Inami, G. B., S. M. C. Lee, R. W. Hogue, and R. A. Brenden.** 2001. Two processing methods for the isolation of *Salmonella* from naturally contaminated alfalfa seeds. *J. Food Prot.* **64:**1240–1243.
86. **International Specialty Supply.** 2 August 2004, posting date. *Comment on Produce Safety from Production to Consumption, Food and Drug Administration Docket No. 2004N-0258.* [Online.] International Specialty Supply, Cookeville, Tenn. http://www.fda.gov/ohrms/dockets/dailys/04/july04/071604/04N-0258–emc00001–01.pdf.
87. **International Specialty Supply.** 2005, posting date. *Seed Sampling for Human Pathogen Testing.* [Online.] International Specialty Supply, Cookeville, Tenn. http://www.sproutnet.com/sprouting_seed_ safety.htm.
88. **International Sprout Growers Association.** 25 October 2002, revision date. *Hazard Analysis Critical Control Point (HACCP) Check List.* [Online.] International Sprout Growers Association, Seattle, Wash. http://www.isga-sprouts.org/haccp.htm.
89. **Itoh, Y., Y. Sugita-Konishi, F. Kasuga, M. Iwaki, Y. Hara-Kudo, N. Saito, Y. Noguchi, H. Konuma, and S. Kumagai.**1998. Enterohemorrhagic *Escherichia coli* O157:H7 present in radish sprouts. *Appl. Environ. Microbiol.* **64:**1532–1535.
90. **Jaquette, C. B., L. R. Beuchat, and B. E. Mahon.** 1996. Efficacy of chlorine and heat treatment in killing *Salmonella stanley* inoculated onto alfalfa seeds and growth and survival of the pathogen during sprouting and storage. *Appl. Environ. Microbiol.* **62:**2212–2215.

91. **Kent, A. D., and E. W. Triplett.** 2002. Microbial communities and their interactions in soil and rhizosphere ecosystems. *Annu. Rev. Microbiol.* **56**:211–236.
92. **Kim, C., Y.-C. Hung, R. E. Brackett, and C.-S. Lin.** 2003. Efficacy of electrolyzed oxidizing water in inactivating *Salmonella* on alfalfa seeds and sprouts. *J. Food Prot.* **66**:208–214.
93. **Kim, H.-Y., D. S. Lee, and H.-D. Paik.** 2004. Characterization of *Bacillus cereus* isolates from raw soybean sprouts. *J. Food Prot.* **67**:1031–1035.
94. **Kramer, M. F., and D. V. Lim.** 2004. A rapid and automated fiber optic-based biosensor assay for the detection of *Salmonella* in spent irrigation water used in the sprouting of sprout seeds. *J. Food Prot.* **67**:46–52.
95. **Lang, M. M., B. H. Ingham, and S. C. Ingham.** 2000. Efficacy of novel organic acid and hypochlorite treatments for eliminating *Escherichia coli* O157:H7 from alfalfa seeds prior to sprouting. *Int. J. Food Microbiol.* **58**:73–82.
96. **Lee, S.-Y., K.-M. Yun, J. Fellman, and D.-H.Kang.** 2002. Inhibition of *Salmonella* Typhimurium and *Listeria monocytogenes* in mung bean sprouts by chemical treatment. *J. Food Prot.* **65**:1088–1092.
97. **Leverentz, B., W. S. Conway, W. Janisiewicz, and M. J. Camp.** 2004. Optimizing concentration and timing of a phage spray application to reduce *Listeria monocytogenes* on honeydew melon tissue. *J. Food Prot.* **67**:1682–1686.
98. **Li, Y., and A. Mustapha.** 2004. Simultaneous detection of *Escherichia coli* O157:H7, *Salmonella*, and *Shigella* in apple cider and produce by a multiplex PCR. *J. Food Prot.* **67**:27–33.
99. **Liao, C.-H., and L. M. Shollenberger.** 2003. Detection of *Salmonella* by indicator agar media and PCR as affected by alfalfa seed homogenates and native bacteria. *Lett. Appl. Microbiol.* **36**:152–156.
100. **Liao, C.-H., and W. F. Fett.** 2003. Isolation of *Salmonella* from alfalfa seed and demonstration of impaired growth of heat-injured cells in seed homogenates. *Int. J. Food Microbiol.* **82**:245–253.
101. **Liming, S. H., and A. A. Bhagwat.** 2004. Application of a molecular-beacon-real-time PCR technology to detect *Salmonella* species contaminating fruits and vegetables. *Int. J. Food Microbiol.* **95**:177–187.
102. **Lin, Y.-Y., and T.-J. Fu.** 2004. Evaluation of a fiber-optic biosensor for detection of *Escherichia coli* O157:H7 in fresh produce. *J. Food Prot.* **67**(Suppl. A):128, paper no. p236.
103. **Lindqvist, R., Y. Andersson, B. de Jong, and P. Norberg.** 2000. A summary of reported foodborne disease incidents in Sweden, 1992 to 1997. *J. Food Prot.* **63**:1315–1320.
104. **Matos, A., J. L. Garland, and W. F. Fett.** 2002. Composition and physiological profiling of sprout-associated microbial communities. *J. Food Prot.* **65**:1903–1908.
105. **Matos, A., and J. Garland.** 2005. The effects of community versus single strain inoculants on the biocontrol of *Salmonella* and microbial community dynamics in alfalfa sprouts. *J. Food Prot.* **68**:40–48.
106. **Maude, R. B.** 1996. *Seedborne Diseases and Their Control: Principles & Practice.* CAB International, Wallingford, Oxford, United Kingdom.
107. **Mazzoni, A. M., R. R. Sharma, A. Demirci, and G. R. Ziegler.** 2001. Supercritical carbon dioxide treatments to inactivate aerobic microorganisms on alfalfa seeds. *J. Food Safety* **21**:215–223.

108. **McMahon, M. A. S., and I. G. Wilson.** 2001. The occurrence of enteric pathogens and *Aeromonas* species in organic vegetables. *Int. J. Food Microbiol.* **70:**155–162.
109. **Michino, H., K. Araki, S. Minami, S. Takaya, N. Sakai, M. Miyazaki, A. Ono, and H. Yanagawa.** 1999. Massive outbreak of *Escherichia coli* O157:H7 infection in school children in Sakai City, Japan, associated with consumption of white radish sprouts. *Am. J. Epidemiol.* **150:**787–796.
110. **Mølbak, L., T. R. Licht, T. Kvist, N. Kroer, and S. R. Andersen.** 2003. Plasmid transfer from *Pseudomonas putida* to the indigenous bacteria on alfalfa sprouts: characterization, direct quantification, and in situ location of transconjugent cells. *Appl. Environ. Microbiol.* **69:**5536–5542.
111. **Monier, J.-M., and S. E. Lindow.** 2003. Differential survival of solitary and aggregated bacterial cells promotes aggregate formation on leaf surfaces. *Proc. Natl. Acad. Sci. USA* **100:**15977–15982.
112. **Montville, R., and D. W. Schaffner.** 2004. Analysis of published sprout seed sanitization studies shows treatments are highly variable. *J. Food Prot.* **67:**758–765.
113. **Montville, R., and D. W. Schaffner.** 2005. Monte Carlo simulation of pathogen behavior during the sprout production process. *Appl. Environ. Microbiol.* **71:**746–753.
114. **Muhammad-Tahir, Z., and E. C. Alocilja.** 2004. A disposable biosensor for pathogen detection in fresh produce samples. *Biosyst. Eng.* **88:**145–151.
115. **Mundt, J. O., and N. F. Hinkle.** 1976. Bacteria within ovules and seeds. *Appl. Environ. Microbiol.* **32:**694–698.
116. **Nandiwada, L. S., G. P. Schamberger, H. W. Schafer, and F. Diez-Gonzalez.** 2004. Characterization of a novel E2-type colicin and its application to treat alfalfa seeds to reduce *Escherichia coli* O157:H7. *Int. J. Food Microbiol.* **93:**267–279.
117. **National Advisory Committee on Microbiological Criteria for Foods.** 1999. Microbiological safety evaluations and recommendations on sprouted seeds. *Int. J. Food Microbiol.* **52:**123–153.
118. **Nisbet, D.** 2002. Defined competitive exclusion cultures in the prevention of enteropathogenic colonization in poultry and swine. *Antonie Leeuwenhoek* **81:**481–486.
119. **Onoue, Y., H. Konuma, H. Nakagawa, Y. Hara-Kudo, T. Fujita, and S. Kumagai.** 1999. Collaborative evaluation of detection methods for *Escherichia coli* O157:H7 from radish sprouts and ground beef. *Int. J. Food Microbiol.* **46:**27–36.
120. **Palmai, M., and R. L. Buchanan.** 2002. Growth of *Listeria monocytogenes* during germination of alfalfa sprouts. *Food Microbiol.* **19:**195–200.
121. **Palmai, M., and R. L. Buchanan.** 2002. The effect of *Lactococcus lactis* on the growth characteristics of *Listeria monocytogenes* in alfalfa sprout broth. *Acta Aliment.* **31:**379–392.
122. **Pangrangi, S., M. W. Elwell, R. C. Anantheswaran, and L. F. LaBorde.** 2003. Efficacy of sulfuric acid scarification and disinfection treatments in eliminating *Escherichia coli* O157:H7 from alfalfa seeds prior to sprouting. *J. Food Sci.* **68:**613–618.
123. **Pao, S., S. P. Randolph, E. W. Westbrook, and H. Shen.** 2004. Use of bacteriophages to control *Salmonella* in experimentally contaminated sprout seeds. *J. Food Sci.* **69:**M127–M130.
124. **Park, C. E., and G. W. Sanders.** 1990. Source of *Klebsiella pneumoniae* in alfalfa and mung bean sprouts and attempts to reduce its occurrence. *Can. Inst. Food Sci. Technol. J.* **23:**189–192.

125. **Park, C. M., P. J. Taormina, and L. R. Beuchat.** 2000. Efficacy of allyl isothiocyanate in killing enterohemorrhagic *Escherichia coli* O157:H7 on alfalfa seeds. *Int. J. Food Microbiol.* **56**:13–20.

126. **Park, W. P., S. H. Cho, and D. S. Lee.** 1998. Effect of minimal processing operations on the quality of garlic, green onion, soybean sprouts and watercress. *J. Sci. Food Agric.* **77**:282–286.

127. **Patterson, J. E., and M. J. Woodburn.** 1980. *Klebsiella* and other bacteria on alfalfa and bean sprouts at the retail level. *J. Food Sci.* **45**:492–495.

128. **Piernas, V., and J. P. Guiraud.** 1997. Microbial hazards related to rice sprouting. *Int. J. Food Sci. Technol.* **32**:33–39.

129. **Piernas, V., and J. P. Guiraud.** 1998. Control of microbial growth on rice sprouts. *Int. J. Food Sci. Technol.* **33**:297–305.

130. **Pönkä, A., Y. Andersson, A. Siitonen, B. de Jong, M. Jahkola, O. Haikala, A. Kuhmonen, and P. Pakkala.** 1995. *Salmonella* in alfalfa sprouts. *Lancet* 345:462–463.

131. **Portnoy, B. L., J. M. Goepfert, and S. M. Harmon.** 1976. An outbreak of *Bacillus cereus* food poisoning resulting from contaminated vegetable sprouts. *Am. J. Epidemiol.* **103**: 589–594.

132. **Price, T. V.** 1988. Seed sprout production for human consumption—a review. *Can. Inst. Food Sci. Technol.* **21**:57–65.

133. **Proctor, M. E., M. Hamacher, M. L. Tortorello, J. R. Archer, and J. P. Davis.** 2001. Multistate outbreak of *Salmonella* serovar Muenchen infections associated with alfalfa sprouts grown from seeds pretreated with calcium hypochlorite. *J. Clin. Microbiol.* **39**:3461–3465.

134. **Prokopowich, D., and G. Blank.** 1991. Microbiological evaluation of vegetable sprouts and seeds. *J. Food Prot.* **54**:560–562.

135. **Puohiniemi, R., T. Heiskanen, and A. Siitonen.** 1997. Molecular epidemiology of two international sprout-borne *Salmonella* outbreaks. *J. Clin. Microbiol.* **35**:2487–2491.

136. **Rajkowski, K. T., and D. W. Thayer.** 2000. Reduction of *Salmonella* spp. and strains of *Escherichia coli* O157:H7 by gamma radiation of inoculated sprouts. *J. Food Prot.* **63**:871–875.

137. **Rajkowski, K. T., and D. W. Thayer.** 2001. Alfalfa seed germination and yield ratio and alfalfa sprout microbial keeping quality following irradiation of seeds and sprouts. *J. Food Prot.* **64**:1988–1995.

138. **Rajkowski, K. T., G. Boyd, and D. W. Thayer.** 2003. Irradiation D-values for *Escherichia coli* O157:H7 and *Salmonella* sp. on inoculated broccoli seeds and effects of irradiation on broccoli sprout keeping quality and seed viability. *J. Food Microbiol.* **66**:760–766.

139. **Robertson, L. J., B. Gjerde, and A. T. Campbell.** 2000. Isolation of *Cyclospora* oocysts from fruits and vegetables using lectin-coated paramagnetic beads. *J. Food Prot.* **63**:1410–1414.

140. **Robertson, L. J., and B. Gjerde.** 2001. Occurrence of parasites on fruits and vegetables in Norway. *J. Food Prot.* **64**:1793–1798.

141. **Robertson, L. J., and B. Gjerde.** 2001. Factors affecting recovery efficiency in isolation of *Cryptosporidium* oocysts and *Giardia* cysts from vegetables for standard method development. *J. Food Prot.* **64**:1799–1805.

142. **Robertson, L. J., G. S. Johannessen, B. K. Gjerde, and S. Loncarevic.** 2002. Microbiological analysis of seed sprouts in Norway. *Int. J. Food Microbiol.* **75**:119–126.

143. **Sanderson, B.** 2004, posting date. *Seed Sampling and Testing: a Risk-Reduction Strategy for Sprouts.* [Online.] International Specialty Supply, Cookeville, Tenn. http://www.sproutnet.com/Research/seed_sampling_and_testing.htm.

144. **Sata, S., T. Fujisawa, R. Osawa, A. Iguchi, S. Yamai, and T. Shimada.** 2003. An improved enrichment broth for isolation of *Escherichia coli* O157, with specific reference to starved cells, from radish sprouts. *Appl. Environ. Microbiol.* **69:**1858–1860.

145. **Schloss, P. D., and J. Handelsman.** 2004. Status of the microbial census. *Microbiol. Mol. Biol. Rev.* **68:**686–691.

146. **Schoeller, N. P., S. C. Ingham, and B. H. Ingham.** 2002. Assessment of the potential for *Listeria monocytogenes* survival and growth during alfalfa sprout production and use of ionizing radiation as a potential intervention treatment. *J. Food Prot.* **65:**1259–1266.

147. **Scouten, A. J., and L. R. Beuchat.** 2002. Combined effects of chemical, heat and ultrasound treatments to kill *Salmonella* and *Escherichia coli* O157:H7 on alfalfa seeds. *J. Appl. Microbiol.* **92:**668–674.

148. **Sewell, A. M., and J. M. Farber.** 2001. Foodborne outbreaks in Canada linked to produce. *J. Food Prot.* **64:**1863–1877.

149. **Sharma, R. R., and A. Demirci.** 2003. Treatment of *Escherichia coli* O157:H7 inoculated alfalfa seeds and sprouts with electrolyzed oxidizing water. *Int. J. Food Microbiol.* **86:**231–237.

150. **Sharma, R. R., A. Demirci, L. R. Beuchat, and W. F. Fett.** 2002. Inactivation of *Escherichia coli* O157:H7 on inoculated alfalfa seeds with ozonated water and heat treatment. *J. Food Prot.* **65:**447–451.

151. **Sharma, R. R., A. Demirci, L. R. Beuchat, and W. F. Fett.** 2002. Inactivation of *Escherichia coli* O157:H7 on inoculated alfalfa seeds with ozonated water under pressure. *J. Food Safety* **22:**107–119.

152. **Shearer, A. E. H., C. M. Strapp, and R. D. Joerger.** 2001. Evaluation of a polymerase chain reaction-based system for detection of *Salmonella* Enteritidis, *Escherichia coli* O157:H7, *Listeria* spp., and *Listeria monocytogenes* on fresh fruits and vegetables. *J. Food Prot.* **64:**788–795.

153. **Singh, N., R. K. Singh, and A. K. Bhunia.** 2003. Sequential disinfection of *Escherichia coli* O157:H7 inoculated alfalfa seeds before and during sprouting using aqueous chlorine dioxide, ozonated water, and thyme essential oil. *LWT Food Sci. Technol.* **36:**235–243.

154. **Sivapalasingam, S., C. R. Friedman, L. Cohen, and R. V. Tauxe.** 2004. Fresh produce: a growing cause of outbreaks of foodborne illness in the United States, 1973 through 1997. *J. Food Prot.* **67:**2342–2353.

155. **Skowronek, F., L. Simon-Sarkadi, and W. H. Holzapfel.** 1998. Hygenic status and biogenic amine content of mung bean sprouts. *Z. Lebensm. Unters. Forsch. A* **207:**97–100.

156. **Sly, T., and E. Ross.** 1982. Chinese foods: relationship between hygiene and bacterial flora. *J. Food Prot.* **45:**115–118.

157. **Smith, M.** 2000. Microbial testing of spent irrigation water during sprout production. *Food Safety Magazine* **6:**8, 46.

158. **Splittstoesser, D. F., D. T. Queale, and B. W. Andaloro.**1983. The microbiology of vegetable sprouts during commercial production. *J. Food Safety* **5:**79–86.

159. **Stan, S. D., and M. A. Daeschel.** 2003. Reduction of *Salmonella enterica* on alfalfa seeds with acidic electrolyzed oxidizing water and enhanced uptake of acidic electrolyzed oxidizing water into seeds by gas exchange. *J. Food Prot.* **66:**2017–2022.

160. **Stewart, D. S., K. Reineke, J. Ulaszek, T. Fu, and M. Tortorello.** 2001. Growth of *Escherichia coli* O157:H7 during sprouting of alfalfa seeds. *Lett. Appl. Microbiol.* **33:**95–99.

161. **Stewart, D. S., K. F. Reineke, J. M. Ulaszek, and M. L. Tortorello.** 2001. Growth of *Salmonella* during sprouting of alfalfa seeds associated with salmonellosis outbreaks. *J. Food Prot.* **64:**618–622.

162. **Stewart, D. S., K. F. Reineke, and M. L. Tortorello.** 2002. Comparison of Assurance Gold *Salmonella* EIA, BAX for Screening/*Salmonella*, and GENE-TRAK *Salmonella* DLP rapid assays for detection of *Salmonella* in alfalfa sprouts and sprout irrigation water. *J. AOAC Int.* **85:**395–403.

163. **Strapp, C. M., A. E. H. Shearer, and R. D. Joerger.** 2003. Survey of retail alfalfa sprouts and mushrooms for the presence of *Escherichia coli* O157:H7, *Salmonella*, and *Listeria* with BAX, and evaluation of this polymerase chain reaction-based system with experimentally contaminated samples. *J. Food Prot.* **66:**182–187.

164. **Suslow, T. V., J. Wu, W. F. Fett, and L. J. Harris.** 2002. Detection and elimination of *Salmonella* Mbandaka from naturally contaminated alfalfa seed by treatment with heat or calcium hypochlorite. *J. Food Prot.* **65:**452–458.

165. **Taitt, C. R., Y. S. Shubin, R. Angel, and F. S. Ligler.** 2004. Detection of *Salmonella enterica* serovar Typhimurium by using a rapid, array-based immunosensor. *Appl. Environ. Microbiol.* **70:**152–158.

166. **Taormina, P. J., and L. R. Beuchat.** 1999. Comparison of chemical treatments to eliminate enterohemorrhagic *Escherichia coli* O157:H7 on alfalfa seeds. *J. Food Prot.* **62:**318–324.

167. **Taormina, P. J., and L. R. Beuchat.** 1999. Behavior of enterohemorrhagic *Escherichia coli* O157:H7 on alfalfa sprouts during the sprouting process as influenced by treatments with various chemicals. *J. Food Prot.* **62:**850–856.

168. **Taormina, P. J., L. R. Beuchat, and L. Slutsker.** 1999. Infections associated with eating seed sprouts: an international concern. *Emerg. Infect. Dis.* **5:**626–634.

169. **Thayer, D. W., G. Boyd, and W. F. Fett.** 2003. γ-Radiation decontamination of alfalfa seeds naturally contaminated with *Salmonella* Mbandaka. *J. Food Sci.* **68:**1777–1781.

170. **Thomas, J. L., M. S. Palumbo, J. A. Farrar, T. B. Farver, and D. O. Cliver.** 2003. Industry practices and compliance with the U.S. Food and Drug Administration Guidelines among California sprout firms. *J. Food Prot.* **66:**1253–1259.

171. **Thomashow, L. S., and D. M. Weller.** 1988. Role of a phenazine antibiotic from *Pseudomonas fluorescens* in biological control of *Gaeumannomyces graminis* var. *tritici*. *J. Bacteriol.* **170:**3499–3508.

172. **Thunberg, R. L., T. T. Tran, R. W. Bennet, R. N. Matthews, and N. Belay.** 2002. Microbial evaluation of selected fresh produce obtained at retail markets. *J. Food Prot.* **65:**677–682.

173. **Tkalcic, S., T. Zhao, B. G. Harmon, M. P. Doyle, C. A. Brown, and P. Zhao.** 2003. Fecal shedding of enterohemorrhagic *Escherichia coli* in weaned calves following treatment with probiotic *Escherichia coli*. *J. Food Prot.* **66:**1184–1189.

174. **Tortorello, M. L., D. S. Stewart, K. F. Reineke, and J. M. Ulaszek.** 2000. *Detection of Salmonella in Alfalfa Sprout Spent Irrigation Water by Immunoassays.* Food and Drug Administration laboratory information bulletin 4214. Food and Drug Administration, Washington, D.C.

175. **Tu, S.-I., M. Golden, W. F. Fett, A. Gehring, and P. Irwin.** 2003. Rapid detection of outbreak *Escherichia coli* O157 and *Salmonella* on alfalfa sprouts by immunomagnetic capture and time-resolved fluorescence. *J. Food Safety* **23**:75–89.

176. **USDA Agricultural Marketing Service.** 3 November 2003, posting date. *National Organic Program.* [Online.] USDA Agricultural Marketing Service, Washington, D.C. http://www.ams.usda.gov/nop/NOP/Standards/ListReg.html.

177. **Viswanathan, P., and R. Kaur.** 2001. Prevalence and growth of pathogens on salad vegetables, fruits and sprouts. *Int. J. Environ. Health* **203**:205–213.

178. **Wade, W. N., A. J. Scouten, K. H. McWatters, R. L. Wick, A. Demirci, W. F. Fett, and L. R. Beuchat.** 2003. Efficacy of ozone in killing *Listeria monocytogenes* on alfalfa seeds and sprouts and effects on sensory quality of sprouts. *J. Food Prot.* **66**:44–51.

179. **Warriner, K., S. Spaniolas, M. Dickinson, C. Wright, and W. M. Waites.** 2003. Internalization of bioluminescent *Escherichia coli* and *Salmonella* Montevideo in growing bean sprouts. *J. Appl. Microbiol.* **95**:719–727.

180. **Warriner, K. S., F. Ibrahim, M. Dickinson, C. Wright, and W. M. Waites.** 2003. Internalization of human pathogens within growing salad vegetables. *Biotechnol. Genet. Eng. Rev.* **20**:117–134.

181. **Watanabe, Y., K. Ozasa, J. H. Mermin, P. M. Griffin, K. Masuda, S. Imashuku, and T. Sawada.** 1999. Factory outbreak of *Escherichia coli* O157:H7 infection in Japan. *Emerg. Infect. Dis.* **5**:424–428.

182. **Weagent, S. D., and A. J. Bound.** 2000. *Comparison of Detection Methods for* Escherichia coli *O157:H7 from Artificially Contaminated Spent Irrigation Water from Sprouted Seeds.* Food and Drug Administration laboratory information bulletin 4205. Food and Drug Administration, Washington, D.C.

183. **Weagant, S. D., and A. J. Bound.** 2001. Evaluation of techniques for enrichment and isolation of *Escherichia coli* O157:H7 from artificially contaminated sprouts. *Int. J. Food Microbiol.* **71**:87–92.

184. **Weiss, A., and W. P. Hammes.** 2003. Thermal seed treatment to improve the food safety status of sprouts. *J. Appl. Bot.* **77**:152–155.

185. **Weissinger, W. R., and L. R. Beuchat.** 2000. Comparison of aqueous chemical treatments to eliminate *Salmonella* on alfalfa seeds. *J. Food Prot.* **63**:1475–1482.

186. **Weissinger, W. R., K. H. McWatters, and L. R. Beuchat.** 2001. Evaluation of volatile chemical treatments for lethality to *Salmonella* on alfalfa seeds and sprouts. *J. Food Prot.* **64**:442–450.

187. **Wilderdyke, M. R., D. A. Smith, and M. M. Brashears.** 2004. Isolation, identification, and selection of lactic acid bacteria from alfalfa sprouts for competitive inhibition of foodborne pathogens. *J. Food Prot.* **67**:947–951.

188. **Winthrop, K. L., M. S. Palumbo, J. A. Farrar, J. C. Mohle-Boetani, S. Abbott, M. E. Beatty, G. Inami, and S. B. Werner.** 2003. Alfalfa sprouts and *Salmonella* Kottbus infection: a multistate outbreak following inadequate seed disinfection with heat and chlorine. *J. Food Prot.* **66**:13–17.

189. **Wu, F. M., L. R. Beuchat, J. G. Wells, L. Slutsker, M. P. Doyle, and B. Swaminathan.** 2001. Factors influencing the detection and enumeration of *Escherichia coli* O157:H7 on alfalfa seeds. *Int. J. Food Microbiol.* **71**:93–99.

190. **Wu, L., M. H. N. Ashraf, M. Facci, R. Wand, P. G. Paterson, A. Ferrie, and B. H. J. Juurlink.** 2004. Dietary approach to attenuate oxidative stress, hypertension, and inflammation in the cardiovascular system. *Proc. Natl. Acad. Sci. USA* **101**:7094–7099.
191. **Wuytack, E. Y., A. M. J. Diels, K. Meersseman, and C. W. Michiels.** 2003. Decontamination of seeds for sprout production by high hydrostatic pressure. *J. Food Prot.* **66**:918–923.
192. **Yang, C.-H., D. E. Crowley, J. Borneman, and N. T. Keen.** 2001. Microbial phyllosphere populations are more complex than previously realized. *Proc. Natl. Acad. Sci. USA* **98**:3889–3894.
193. **Zhao, T., M. R. S. Clavero, M. P. Doyle, and L. R. Beuchat.** 1997. Health relevance of the presence of fecal coliforms in iced tea and leaf tea. *J. Food Prot.* **60**:215–218.

Microbiology of Fresh Produce
Edited by Karl R. Matthews

Consumer Handling of Fresh Produce

7

Christine M. Bruhn

Consumers believe that eating fresh produce is healthy, and many say that they are increasing their consumption. Most food safety concerns that consumers associate with produce consumption are pesticide related, with few associating the eating of fruits and vegetables with harmful microbes. When problems are detected, consumption drops, but people return to previous consumption patterns after the publicity around a contamination incident subsides. While some consumers carefully wash all produce items, many believe that produce is already clean and that further washing is unnecessary. The produce market is well situated for further expansion when good-flavored products with added convenience are offered. Potential concerns may arise from consumer mishandling, with negative publicity associated with an entire commodity line.

PRODUCE SELECTION

The Food Marketing Institute's annual survey of 1,000 households consistently indicates that good taste is the most important factor influencing purchase of food items (47). Convenience is also considered a very important factor in food selection. Healthfulness and nutritional value are important components, with price considered by fewer people to be very important (48). Additionally, high-quality produce is second only to a clean, neat appearance as the top factor influencing selection of a supermarket. From 1992 to 2004, 90% of consumers or more rated high-quality produce as very important in supermarket selection (48). High-quality fruits and vegetables

Christine M. Bruhn, Center for Consumer Research, University of California, Davis, Davis, CA 95616-8598.

were rated as very important by all income and geographic groups but were especially valued among the highest-income households, where 97% rated them very important in 2000 (47).

Consumers view produce as a healthy food choice. In each of the last 10 years, 70% or more consumers responding to the Food Marketing Institute's annual survey have said that they have increased produce consumption for a healthier diet (48). Similarly, Americans responding to *Parade*'s annual study of the nation's shopping practices say that they are eating more vegetables (50%), salads (49%), and fruits (47%) (22). Consumers view fruits and vegetables as good sources of vitamins, minerals, and fiber and as foods that are helpful in calorie control and possible cancer preventatives. Surveys of focus groups conducted over the Internet by Vance Research Services found that consumers were aware of the link between produce and cancer prevention (35). In 1997, Californians said that they were eating more fruits and vegetables because they were "trying to eat healthier" (30%), "liking the taste" (30%), "lowering disease risk" (6%), and aiming for "weight reduction" (5%) (14). When asked in 2004 what health-related reasons made them eat more produce, people responded similarly; they were cutting back on calories, reducing cholesterol, following a diet, or following suggestions from a health professional (35). More households with children said that they were increasing their produce consumption than did households without children. Apples, bananas, carrots, bagged salads, and broccoli are the top five fruits and vegetables that consumers reported eating more of in 2004 (35). Marketplace disappearance data confirm an increase in fruit and vegetable consumption, with the greatest increase in the fresh compared to the frozen or canned category (13).

Consumers choose produce based upon product appearance, associating good taste and optimal nutritional value with color and freshness (19, 41). Although color varies by produce and variety, red blush is preferred in some products, like peaches and nectarines (10). A characteristic odor is desirable as it indicates ripeness and reflects eating quality. Generally, larger products are priced at a premium; however, some consumers prefer medium or small products depending on intended use. Scars, scratches, and other marks lower quality ratings (29); however, some consumers will purchase lower grades if the price is sufficiently low and other factors indicate good eating quality. Consumers avoid produce with cuts, bruises, or obvious blemishes. In a 2005 survey, 33% of consumers reported discovering that the produce they purchased in the past year was bruised or damaged (45). While some blemishes are cosmetic, avoiding products in which the tissue is broken protects safety as bacteria have been shown to spread more rapidly in fruit with cuts and bruises.

Convenience is highly valued among today's consumers (21). Consumers report an increased use of pretrimmed, -washed, and -bagged fresh produce. Overall, 32% indicated that they purchase convenience products more frequently now than they did 5 years ago (44). More households with children purchased convenience products (31%) than households without children (25%).

Some consumers do know how to choose and how to prepare some new produce items from the supermarket. While lettuce and broccoli are familiar to most, fennel, mango, and other specialty products are new to many consumers. In 1999, 45% of consumers reported purchasing leafy green vegetables, including salad mixes, for the first time and 14% of consumers indicated that they had tried a new vegetable for the first time (40). When allowed to use food stamps for fresh produce at the farmer's market, 56% of consumers report trying items for the first time. This suggests that low-income consumers may also be unfamiliar with proper selection and safe handling practices for new items (28). Attitude studies indicate that consumers tend to prefer locally grown produce, due both to perceptions of higher quality and to a desire to support the local economy; however, many people do not know what produce is grown locally (7, 32).

Consumer recommendations on produce promotion suggest that supermarkets do not always follow recommended practices. In a 2005 survey, 10% of consumers reported that they stopped buying specific items because of concern about the cleanliness of the store or produce department (45). Consumers suggest that spoiled produce be removed from the display, products be held at proper temperatures, and recipes and preparation tips be offered (37, 38).

CONSUMER PERCEPTION OF PRODUCE SAFETY

Most consumers are confident in the safety of the food supply; however, perception varies depending on news coverage of food safety issues. The Food Market Institute's annual survey indicated that in 2005, 85% of consumers were completely or mostly confident that food in the supermarket was safe (16). When asked where they thought food safety problems were most likely to occur, 30% cited the food-processing or -manufacturing plant, 20% indicated restaurants, and 18% responded the home. Only 1% indicated that food safety problems were likely to occur on the farm. In 2005, 11% of consumers said that they stopped buying specific food products because of food safety concerns. When asked what food they stopped purchasing, 53% indicated meat or poultry, 13% indicated seafood, and 12% indicated fruits and vegetables. Green onions, or scallions, were identified by 6% of consumers as a food that they no longer buy and strawberries were identified by 2% (16).

When food safety in general is considered, bacterial contamination is identified as a serious health risk by the largest percentage of consumers, 58% (16). Concern about pesticide residues is mentioned by 47% of consumers. In contrast, when consumers are asked about safety concerns associated with produce, pesticide residues are mentioned most frequently (42). In 2001, 62% of consumers mentioned concerns associated with fruits and 41% expressed concern about vegetables. Apples, grapes, and strawberries are more frequently listed as being prone to food safety concerns than other fruits, while lettuce leads the list of vegetables with potential concerns.

Perception of Organic Produce

In response to concern about pesticide residues, many supermarkets offer organic produce. The organic market grew from a sales level of $2.3 billion in 1994 to one of $9.7 billion in 2002. Organic foods are found in food service operations and mainstay supermarkets as well as specialized markets. In 2003, 51% of U.S. women indicated that they had seen the U.S. Department of Agriculture organic seal at the store where they did most of their shopping (11). The largest organic market is that for fresh produce (36). Correspondingly, 92% of organic-product users have purchased fruits and vegetables (23).

The Hartman Group (23) found that consumers select organic products for health and nutrition reasons, followed by taste, belief in food safety, and environmental concerns. These consumers believe that consuming conventionally produced food is risky. They select organic products because they believe that these products are grown without pesticides, are chemical free, and are safer for the environment (23, 24, 34, 51). Eating organic food is mentioned as a practice used to maintain health by 37% of consumers (17). The statement "grown without pesticides" has been rated as very important to more consumers than the phrase "certified organic" (24).

Perception of Produce Grown under IPM

Virtually every land grant university in the United States has a research group developing environmentally responsive pest control strategies as part of an integrated pest management (IPM) approach. These methods include use of good insects to attack harmful ones, use of insect-resistant varieties of plants, and use of production management techniques. As a last resort, if pests reach an economically significant level, pesticides may be employed. Prior to selection of a pesticide, the impacts on workers, the environment, and the target pest are evaluated.

When consumers hear about the IPM approach, their attitudes toward farming practices and food safety are positive (8). Govindasamy and colleagues (20) found that people supported IPM through both willingness to purchase and willingness to pay a premium for produce grown under IPM. Once informed about IPM, more consumers were willing to switch supermarkets and pay a premium to obtain IPM rather than organic produce. An economic analysis of purchase intent found that those with higher-than-average incomes, young individuals, those who frequently purchase organic produce, and those who live in suburban areas are likely to purchase produce grown under IPM and to pay a premium for it (20).

MICROBIOLOGICAL HAZARDS

Consumers say that the greatest threat to food safety is microbiological (12, 48). Outbreaks of foodborne illness have occurred in which fresh produce was identified as the source of the pathogen. Cantaloupes were the source of infections with *Salmonella* spp., frozen strawberries were contaminated with hepatitis A virus, and lettuce, sprouts, fresh apple juice, and fresh basil were implicated in outbreaks of *Escherichia coli* O157:H7 infection. Consumers responded by avoiding the implicated products. After these incidences, as many as 60% of consumers indicated that they were more concerned about bacterial contamination of fresh produce than in the previous year (39). Branded produce is considered somewhat or much more likely to be safe than nonbranded produce by 32% of consumers (43). While consumers believe that produce grown in the United States is safer than imported produce, the USDA Economic Research Service notes that outbreaks occur from domestic as well as imported produce (52).

CONSUMER HANDLING PRACTICES

Studies of consumer attitudes and self-reported behavior indicate that most people handle foods safely; however, members of every demographic group report mishandling which can result in increased likelihood of foodborne illness (46). Generally, a larger percentage of people over 45 years of age than of those younger than 45 years report following safe handling practices (1, 2, 27, 30, 31, 49), men are less likely than women to follow kitchen sanitation procedures (1, 2, 27, 31), and those with at least some college are less likely to follow safe handling guidelines than those with 12 years or less of schooling (2, 31).

Traditionally, consumer food safety research is based upon self-reported behavior. Recent approaches based upon observation found that people overstate their compliance with safe handling guidelines. For example, Audits International (6) found that 79% of consumers correctly identified instances in which hand washing was necessary during food preparation; however, 29% were observed failing to wash their hands when they should have. Similarly, 97% of consumers believed that eating lettuce that had been moistened by raw poultry drippings was a "risky" food-handling practice, yet 98% of these consumers were observed cross-contaminating ready-to-eat foods with raw meat or raw egg during food preparation (5).

Similarly, a video study of food-handling practices in Australia found significant variance between stated answers on a survey and actual food-handling practices (26). Infrequent hand washing, poor hand-washing techniques, inadequate cleaning of kitchen surfaces, involvement of pets in the kitchen, and frequent touching of the face, mouth, nose, and/or hair occurred during food preparation.

SELECTING PRODUCE AND BRINGING IT HOME

Consumer food safety guidelines emphasize the need to separate raw meat and poultry from foods to be eaten raw; however, many consumers do not realize that the potential for juices to cross-contaminate can occur in the shopping cart or grocery bag. While some consumers routinely separate meat and poultry from raw produce when shopping, focus group discussions indicate that this precaution does not occur to others (31). Less than 30% of consumers polled in a nationwide mailing indicated that they ask for meat, poultry, and fish to be bagged separately from fresh produce (31). More than half of consumers surveyed indicated that they had no special requirements for produce packaging.

Home Storage

Most consumers store produce in the refrigerator; however, some items are stored at room temperature (31). Room temperature storage is appropriate for optimum quality of tomatoes, bananas, and unripe climacteric fruits, but refrigeration lengthens the freshness and slows bacterial growth should a produce item contain harmful microorganisms. According to a nationwide survey, 42% store apples and 24% store melons at room temperature (31).

Where consumers place produce in the refrigerator can affect the potential for cross-contamination. Most consumers store fresh produce either in the refrigerator produce drawer or on a shelf. Produce can be contaminated by meat or poultry juices if stored beneath these foods. In a nationwide survey, 30% of consumers reported placing fruits or vegetables wherever there was room or storing them on a shelf below meat and poultry (31). When specifically asked, only 51% of consumers responded that they keep foods separate to avoid cross-contamination (16).

Contaminants can transfer from unclean surfaces to produce in the refrigerator. Frequent cleaning of the home refrigerator is not a universal practice. While 50% of consumers indicate that they clean their refrigerators at least once a month, the remainder clean two to three times a year or less frequently (31).

Preparation

Although people should wash their hands before beginning food preparation, consumers do not always follow this practice. Between 20 and 74% of consumers reported not washing their hands before starting meal preparation or after handling raw meat or poultry (3, 4, 16, 18, 50). Almost half of the consumers responding to a nationwide survey acknowledged that they do not always wash their hands before handling produce (31). When actual food handling was observed, only 45% of consumers attempted to wash their hands before food preparation; of those who washed, 84% used soap (5). Similarly, nine observational studies of United Kingdom consumers found that 93 to 100% failed to wash and dry their hands adequately on at least one occasion while handling raw meat or poultry (46).

The kitchen can serve as a source of contamination. Slightly more than half of consumers report washing the sink before handling fresh produce and about half wash the sink after handling produce (31). Most, 69%, indicate that they use a cleanser or cleaning solution for washing, 40% use dishwashing liquid, 27% use bleach, 19% use antibacterial soap, and 11% indicate that they wash with water only. Further, between 20 and 70% report using the same utensils and/or cutting board without washing between preparing raw meat or poultry and preparing ready-to-eat raw produce (3, 4, 9, 18, 27, 30, 31, 49, 50).

Consumers report that they wash produce, with 81% saying that they wash just before eating or cooking and 21% saying that they wash before placing produce in the refrigerator; 6% acknowledge that they seldom or never wash produce (31). The item washed least frequently is melon, with 36% indicating they never wash this fruit (31). Many believe it is not necessary to wash melon since the rind is not consumed (31). Consumers have indicated that they did not think it was necessary to wash home-grown or organic produce (31). In actual practice, consumers may not wash produce, even when it will be eaten raw. Anderson et al. (5) observed that vegetable washing was inadequate. When preparing salad, 6 of 99 subjects made no effort to clean the vegetables, 70 rinsed the lettuce, 93 rinsed the tomatoes, 47 rinsed the carrots, and 55 rinsed the cucumbers. Rinsing time ranged from 1 s for tomatoes, cucumbers, and carrots to up to 55 s for tomatoes but averaged less than 12 s.

Consumers use a variety of methods to wash produce. Relatively efficient methods, such as peeling, rubbing with hands, scrubbing with a brush, and washing under running water, have all been reported, with washing under running water being the most common method (31). As many as 20%, however, soak produce in a container, a method not recommended because contamination can be spread to other produce items. As many as 4% wash produce with dish detergent. The Food and Drug Administration does not recommend this practice because residue can remain on the product (15, 18).

Storage of cut fruit can permit the growth of pathogens. In focus group discussions, most consumers recognized that cut fruit should be refrigerated; however, some volunteered that they stored cut melons at room temperature. When commenting on the development of a safe handling practices brochure, consumers advised that the word "always" be added to the advice to refrigerate leftovers, since the participants knew of people who stored melons and other cut fruit at room temperature (31).

FUTURE CONCERNS

In a review of 88 consumer food safety studies from the United States, Europe, Canada, Australia, and New Zealand, Redmond and Griffith observed that the majority of consumers think themselves to be adequately informed regarding food safety (46). They are aware of some safe food-handling practices and unaware of others. Consumer knowledge of safe handling is inadequate and requires improvement. This inadequate knowledge may lead to unsafe food preparation practices and contribute to foodborne disease. Since consumers do not recognize that they need information, sharing safe handling facts and motivating behavior change will be challenging.

Convincing consumers to change unsafe handling practices may be difficult because fruits and vegetables are not commonly associated with foodborne illness. About one-third of those responding to a mail questionnaire on safe handling indicated that they were not interested in receiving information on safe handling practices (31). Therefore, it is important to acknowledge both that fruits and vegetables promote health and that they must be handled appropriately to avoid illness.

Many consumers want to enjoy seasonal produce year round. Fruits and vegetables from other regions of the world are becoming popular. Meeting the demand for year-round fresh produce will stimulate an increase in the number of imported items. Agricultural practices will need to be adopted worldwide that reduce the likelihood of foodborne pathogens on raw products (25).

Cultivation of school vegetable gardens is a popular activity to acquaint children with food production and encourage consumption of fruits and vegetables (33). Garden projects should address planting to prevent the introduction of harmful bacteria, appropriate washing of produce, and proper hand washing, especially before eating.

Consumers will likely respond positively to new partly prepared produce items that feature added convenience. Since refrigeration of cut produce is not practiced by all consumers, advice to keep partly prepared items refrigerated should be printed prominently on the packaging.

REFERENCES

1. **Albrecht, J. A.** 1995. Food safety knowledge and practices of consumers in the U.S.A. *J. Consumer Studies Home Economics* **19**:119–134.
2. **Altekruse, S. F., D. A. Street, S. B. Fein, and A. S. Levy.** 1996. Consumer knowledge of foodborne microbial hazards and food-handling practices. *J. Food Prot.* **59**:287–294.
3. **Altekruse, S. F., S. Yang, B. B.Timbo, and F. J. Angulo.** 1999. A multi-state survey of consumer food handling and food-consumption practices. *Am J. Prev. Med.* **16**:216–221.
5. **Anderson, J. B., T. A. Shuster, K. E. Hansen, A. S. Levy, and A. Vok.** 2004. A camera's view of consumer food-handling behaviors. *Am. Diet. Assoc.* **104**:186–191.
6. **Audits International.** *Home Food Safety Survey Report.* Available from Richard W. Daniels, Rutgers University, by e-mail at info@audits.com.
7. **Bruhn, C.** 1992. Consumer attitude toward locally grown produce. *Calif. Agric.* **46**:13.
8. **Bruhn, C., S. Peterson, P. Phillips, and N. Sakovich.** 1992. Consumer response to information on integrated pest management. *J. Food Safety* **12**:315–326.
9. **Bruhn, C., and H. Schutz.** 1999. Consumer food safety knowledge and practices. *J. Food Safety* **19**:73–87.
10. **Bruhn, C. M.** 1995. Consumer and retailer satisfaction with the quality and size of California peaches and nectarines. *J. Food Quality* **18**:241–256.
11. **Burfields, T.** 15 September 2003, posting date. Customers accepting organics. *Packer Online* [Online.] http://www.thepacker.com/icms/_dtaa2/content/2003–16620–290.asp.
12. **Cogent Research.** June 2005, posting date. [Online.] *Food Biotechnology Not a Top-of-Mind Concern for American Consumers.* International Food Information Center. http://www.ific.org/research/biotechres03.cfm.
13. **Economic Research Service.** Accessed October 2005. *Food Consumption.* [Online.] USDA Economic Research Service, Washington, D.C. http://www.ers.usda.gov/data/foodconsumption/foodavailIndex.htm (custom queries—fruit and vegetable consumption, all years).
14. **Foerster, S. B., J. Gregson, S. Wu, and M. Hudes.** 1998. *California Dietary Practices Survey: Focus on Fruits and Vegetables Trends among Adults, 1989–1997.* California Department of Health Services Public Health Institute, Sacramento, Calif.

15. **Food and Drug Administration.** 26 May 2000, posting date. *FDA Advises Consumers about Fresh Produce Safety.* [Online.] Food and Drug Administration, Washington, D.C. http://vm.cfsan.fda.gov/~lrd/tpproduce.html.
16. **Food Marketing Institute.** 2005. *U.S. Grocery Shopper Trends.* Food Marketing Institute, Washington, D.C.
17. **Food Marketing Institute/Prevention.** 2001. *Shopping for Health 2001: Reaching Out to the Whole Health Consumer.* Food Marketing Institute, Washington, D.C.
18. **Food Safety Inspection Staff, Food Safety and Inspection Service, U.S. Department of Agriculture.** 2000. Focus groups shed light on consumers' food safety knowledge. *Food Safety Educator* **5**:1–8.
19. **Govindasamy, R., J. Italia, and C. Liptak.** 2004, posting date. *Quality of Agricultural Produce: Consumer Preferences and Perceptions.* [Online.] Rutgers University, New Brunswick, N.J. http://aesop.rutgers.edu/%7Eagecon/pub/agmkt.htm. Publication P-02137-1-97, February 1997.
20. **Govindasamy, R., and J. Italia.** 2004, posting date. *Consumer Response to Integrated Pest Management and Organic Agriculture: an Econometric Analysis.* [Online.] Rutgers University, New Brunswick, N.J. http://aesop.rutgers.edu/%7Eagecon/pub/agmkt.htm. Publication P-02137-2-97, November 1997.
21. **Grocery Manufacturers of America.** 2004, posting date. *Convenience Benefits Continue To Make Living Easier.* [Online.] Grocery Manufacturers of America. http://www.gmabrands.org/publications/gmairi/2004/may.htm. *Times and Trends*, p. 12, May 2004.
22. **Hales, D.** 2004. What America really eats. *Sacramento Bee* **14**(Nov.):6–7.
23. **Hartman Group.** 2001. *Healthy Living: Organic and Natural Products Organic Lifestyle Study,* spring. The Hartman Group, Bellevue, Wash.
24. **HealthFocus.** 2003. *HealthFocus Trends Survey.* Health Focus, Atlanta, Ga.
25. **Hedberg, C. W.** 2000. Global surveillance needed to prevent foodborne disease. *Calif. Agric.* **54**:54–61.
26. **Jay, L. S., D. Comar, and L. D. Govenlock.** 1999. A video study of Australian domestic food-handling practices. *J. Food Prot.* **62**:1285–1296.
27. **Jay, L. S., D. Comar, and L. D. Govenlock.** 1999. A national Australian food safety telephone survey. *J. Food Prot.* **62**:921–928.
28. **Joy, A. B., S. Bunch, M. Davis, and J. Fujii.** 2001. USDA program stimulates interest in farmers' markets among low-income women. *Calif. Agric.* **55**:38–41.
29. **Kader, A. A.** 2002. Quality and safety factors: definition and evaluation for fresh horticultural crops, p. 279–285. *In* A. A. Kader (ed.), *Postharvest Technology of Horticultural Crops,* 3rd ed. University of California Agricultural and Natural Resources, Oakland, Calif.
30. **Klontz, K. C., B. Timbo, S. B. Fein, and A. S. Levy.** 1995. Prevalence of selected food consumption and preparation behaviors associated with increased risks of food-borne disease. *J. Food Prot.* **58**:927–930.
31. **Li-Cohen, A. E., and C. M. Bruhn.** 2002. Safety of consumer handling of fresh produce from the time of purchase to the plate: a comprehensive consumer survey. *J. Food Prot.* **65**:1287–1296.
32. **Lockeretz, W.** 1986. Urban consumers' attitudes toward locally grown produce. *Am. J. Altern. Agric.* **1**:83–88.

33. **Morris, J. L., A. Neustadter, and S. Zidenberg-Cherr.** 2001. First grade gardeners more likely to taste vegetables. *Calif. Agric.* **55**:43–46.
34. **Natural Marketing Institute.** 2001. *Health and WellnessTrends Report.* Natural Marketing Institute, Harleysville, Pa.
35. **Nelson, A.** 15 November 2004, posting date. Nutrition news incites consumer purchase. *Packer Online.* [Online.] http://www.thepacker.com/icms/_dtaa2/content/wrapper.asp?alink=2004-1411.2004.
36. **Nutrition Business Journal.** February 2001. The U.S. organic industry. [Online.] http://www.nutritionbusiness.com/index.cfm.
37. **The Packer.** 1996. Catching the shopper's eye. *Fresh Trends* **102**:50–52.
38. **The Packer.** 1996. Turning consumers off. *Fresh Trends* **102**:84–86.
39. **The Packer.** 1998. Microbes grab the spotlight. *Fresh Trends* **104**:20–26.
40. **The Packer.** 1999. Consumers trying greens, salads. *Fresh Trends* **105**:82.
41. **The Packer.** 2001. Eye appeal influences shoppers. *Fresh Trends* **107**:38–42.
42. **The Packer.** 2001. Fruits of concern, igniting fears, food safety first. *Fresh Trends* **2001**:62–65.
43. **The Packer.** 2002. All bets on brands. *Fresh Trends* **108**:11–15.
44. **The Packer.** 2004. By the numbers. *Fresh Trends* **2004**:7–14.
45. **The Packer.** 2005. What specific problem did you have with fresh produce? *Fresh Trends* **2005**:237–241.
46. **Redmond, E. C., and C. J. Griffith.** 2003. Consumer food handling in the home: a review of food safety studies. *J. Food Prot.* **66**:130–161.
47. **Research International.** 2000. *Trends in the Supermarket.* Food Marketing Institute, Washington, D.C.
48. **Research International.** 2004. *Trends in Consumer Attitudes and the Supermarket.* Food Marketing Institute, Washington, D.C.
49. **Williamson, D., R. Gravani, and H. Lawless.** 1992. Correlating food safety knowledge with home food-preparation practices. *Food Technol.* **46**:94–100.
50. **Yang, S., M. G. Leff, D. McTauge, et al.** 1998. Multistate surveillance for food-handling, preparation, and consumption behaviors associated with foodborne diseases: 1995 and 1996 BRFSS food-safety questions. *Morb. Mortal. Wkly. Rep.* **47SS**:33–57.
51. **Zehnder, G., C. Hope, H. Hill, L. Hoyle, and J. Blake.** 2003. An assessment of consumer preferences for IPM and organically grown produce. *J. Ext.* **41**:1–5. [Online.] http://www.joe.org/joe/2003april/rb3.shtml.
52. **Zepp, G., F. Kuchler, and G. Lucier.** 1998. Food safety and fresh fruits and vegetables: is there a difference between imported and domestically produced products? *Vegetables Specialties* **VGS-274**(Apr.):23–28.

Index

S